# Space Science and
# Public Engagement

# Space Science and Public Engagement

## 21st Century Perspectives and Opportunities

*Edited by*

**Amy Paige Kaminski**

ELSEVIER

Elsevier
Radarweg 29, PO Box 211, 1000 AE Amsterdam, Netherlands
The Boulevard, Langford Lane, Kidlington, Oxford OX5 1GB, United Kingdom
50 Hampshire Street, 5th Floor, Cambridge, MA 02139, United States

**Notices**
Knowledge and best practice in this field are constantly changing. As new research and
experience broaden our understanding, changes in research methods, professional
practices, or medical treatment may become necessary.

Practitioners and researchers must always rely on their own experience and knowledge in
evaluating and using any information, methods, compounds, or experiments described
herein. In using such information or methods they should be mindful of their own safety
and the safety of others, including parties for whom they have a professional
responsibility.

To the fullest extent of the law, neither the Publisher nor the authors, contributors, or
editors, assume any liability for any injury and/or damage to persons or property as a
matter of products liability, negligence or otherwise, or from any use or operation of any
methods, products, instructions, or ideas contained in the material herein.

**Library of Congress Cataloging-in-Publication Data**
A catalog record for this book is available from the Library of Congress

**British Library Cataloguing-in-Publication Data**
A catalogue record for this book is available from the British Library

ISBN: 978-0-12-817390-9

For information on all Elsevier publications visit our website at
https://www.elsevier.com/books-and-journals

*Publisher:* Candice Janco
*Acquisitions Editor:* Peter J. Llewellyn
*Editorial Project Manager:* Ruby Smith
*Production Project Manager:* Kiruthika Govindaraju
*Cover Designer:* Alan Studholme

Working together
to grow libraries in
developing countries

www.elsevier.com • www.bookaid.org

Typeset by TNQ Technologies

# Contents

# About the Authors

**Linda Billings** is a consultant to NASA's astrobiology and planetary defense programs in the Planetary Science Division of the Science Mission Directorate at NASA Headquarters in Washington, D.C. She also is director of communications with the Center for Integrative STEM Education at the National Institute of Aerospace in Hampton, Virginia. She lives in Sarasota, Florida. Dr. Billings earned her PhD in mass communication from Indiana University. Her research interests include science and risk communication, social studies of science, and the history and rhetoric of science and space exploration. She has worked for more than 30 years in Washington, D.C., as a researcher; communication planner, manager, and analyst; policy analyst; journalist; and consultant to the government.

**Anton Binneman** is a social scientist with a PhD in narrative. He is currently employed by the National Research Foundation/South African Radio Astronomy Observatory (NRF|SARAO) where he serves as the stakeholder manager for SARAO and the Square Kilometer Array in South Africa. He also is a research associate for the University of Johannesburg, Department of Strategic Communication. Currently, he specializes in strategic stakeholder engagement in science and science infrastructure projects and has more than 10 years' experience in strategic stakeholder communication and engagement, communication research, and public relations management.

**Don Boonstra** has over 40 years of experience as a master STEM educator. Don has been a NASA education specialist, coordinator of the NASA Student Observation Network, and lead for professional development and thematic approaches for NASA planetary science education and public outreach. He is currently a consultant who supports the National Science Teaching Association's online programs. Leveraging his long-term educational experience, Mr. Boonstra has designed innumerable courses to include more innovative methodology, especially the use of computers for data acquisition and manipulation and communication. He has led education teams to develop programs using educational research, best practices, and alignment to the Next Generation Science Standards, including NASA's Mars Student Imaging Project.

**Sanlyn Buxner** is a senior research scientist and senior education and communications specialist at the Planetary Science Institute. She is also an associate research professor of science education in the Department of Teaching, Learning, and Sociocultural Studies at the University of Arizona. Her research interests include quantitative and scientific literacy, the impact of research experiences for students and teachers, and innovative ways to support scientists who participate in public engagement. She publishes broadly in astronomy education research and collaborates on studies of students' knowledge and motivation, research experiences in STEM settings, and outcomes in free-choice learning environments. She has worked in multiple planetariums and outdoor education schools and as an education and outreach specialist for NASA missions. She is the current education and outreach officer for the AAS Division for Planetary Sciences and serves on education committees for both the American Astronomical Society and the American Geophysical Union.

**Corné Davis** is a senior lecturer in the Department of Strategic Communication at the University of Johannesburg. For her PhD, she specialized in cybernetics, specifically second-order cybernetics, and Luhmann's social systems theory, which has become well known since its publication in English in 1995. She has been lecturing on various communication subjects, focusing on strategic communication at both undergraduate and postgraduate levels over the past 12 years and has supervised several research essays and theses for master's degrees. She started working with Anton Binneman and the Square Kilometer Array in October 2018 when he sought to collaborate with the Department of Strategic Communication at the University of Johannesburg.

**Michael Fitzgerald** is a senior research fellow in the School of Education at Edith Cowan University in Perth, Australia. His research interests encompass STEM education, with a focus on astronomy education, as well as pure astronomy research. He has a particular interest in encouraging the use of remotely accessible telescopes to support authentic research and educational activities in the classroom as well as professional learning for high school teachers. He facilitates the yearly Robotic Telescopes, Student Research, and Education Conference connecting amateurs, professionals, and educators with the aim of improving student and teacher outcomes. More broadly he works on research into systemic issues that inhibit and promote STEM education with a variety of foci, including adequate teacher professional development, remote and rural contexts, and student self-efficacy and attitudes. He holds many leadership roles in astronomy education worldwide, including as the current secretary of the International Astronomical Union's Commission C1 for Education and Development.

**Lucy Fortson** is a professor of physics in the School of Physics and Astronomy at the University of Minnesota (UMN). As an observational astrophysicist, she uses very-high-energy gamma ray telescopes to study active galactic nuclei. She is also a founding member of the Zooniverse platform, where over 2 million volunteers contribute to discovery research by performing simple data analysis tasks. Prior to joining the faculty at UMN, Fortson was vice president for research at the Adler Planetarium in Chicago and a research scientist at the University of Chicago. She received her BA from Smith College in physics and astronomy and her PhD from UCLA in high energy physics. She has served on numerous committees including the National Academy of Sciences' 2010 decadal survey for astronomy. She is a Fellow of the American Physical Society, and a member of the American Astronomical Society, the Citizen Science Association and the Association for Computing Machinery. Her awards include most recently the American Physical Society's Nicholson Medal for Outreach.

**Rachel Freed** is a cofounder and the president of the Institute for Student Astronomical Research, having a goal of incorporating scientific research into secondary and undergraduate education. She is currently working on a PhD in astronomy education. She is also a faculty lecturer in the School of Education at Sonoma State University, with a BS degree in biology and an MS in neuroscience. She taught high-school chemistry and astronomy for 10 years and has conducted research on chemistry education, helping to design, build, and evaluate an online formative assessment system for high-school chemistry. She has been an amateur astronomer for more than 20 years and is involved in public outreach. Ms. Freed's work focuses on promoting changes in education that build on students' intrinsic motivations and interests.

**Louis D. Friedman** cofounded The Planetary Society in 1980 with Carl Sagan and Bruce Murray. He served as executive director of the Society for 30 years and remained on the board of directors until 2014. While at the Society, Friedman worked on the Mars Balloon, Mars rover tests, and the Mars microphone projects. He led the Cosmos 1 solar sail mission, and he led the design and development of the LightSail spacecraft. Before his tenure at The Planetary Society, Friedman worked on deep space missions at the Jet Propulsion Laboratory in Pasadena, California. He received a BS in applied mathematics and engineering physics at the University of Wisconsin in 1961, an MS in engineering mechanics at Cornell University in 1963, and a PhD from the aeronautics and astronautics department at the Massachusetts Institute of Technology in 1971.

**Sheri Klug Boonstra** is the director of the Mars Education Program within Arizona State University's School of Earth and Space Exploration and is the principal investigator for the NASA L'SPACE program, a national workforce development opportunity for STEM undergraduate students in the United States. She has worked on NASA mission teams and NASA education projects for over two decades, connecting NASA's exploration of the solar system to immersive, standards-aligned authentic science opportunities for K-12 and higher education students. She has been the formal education lead for NASA's Mars Public Engagement Program and has served as the education and public outreach representative on the Solar System Exploration Subcommittee of the Space Science Advisory Committee for NASA Headquarters. She was the principal investigator for NASA's Undergraduate Student Research Program and has classroom experience as a K-12 science teacher. Ms. Klug Boonstra has received the Excellence in Earth and Space Science

Education Award from the American Geophysical Union in addition to several NASA group achievement awards for her participation and efforts in NASA missions and educational programs.

**Zolt Levay** is retired principal science visuals developer in the Office of Public Outreach at the Space Telescope Science Institute (STScI). He produced images and other visuals from data provided by the Hubble Space Telescope to publicize science results from Hubble and other observatories. He continued this effort in planning for the upcoming James Webb Space Telescope. Mr. Levay also led STScI's Hubble Heritage Team, whose project showcased the visually finest images from Hubble. He became interested in astronomy and photography at an early age and earned a BS in astrophysics at Indiana University Bloomington and an MS in astronomy at Case Western Reserve University. He worked with several space science missions at NASA's Goddard Space Flight Center before joining STScI in 1983.

**Ron Miller** is an author and illustrator specializing in science and astronomy. The author of more than 60 books for both adults and young adults, he is also a regular contributor to magazines such as *Astronomy* and *Scientific American.*

**Mart Noorma** is the science and development director of Milrem Robotics and professor of space and defense technology at the University of Tartu, Estonia. He studied physics at the University of Tartu and received his PhD from Aalto University in 2005. He has worked in many R&D projects related to metrology, space, and defense technologies at Aalto, the National Institute of Standards and Technology in the United States, and Tartu University, where he also served as the vice-rector for academic affairs in 2015–17. He has contributed to many educational initiatives to support and develop teaching quality.

**Arko Olesk** is a lecturer in science communication at Tallinn University, Estonia, where his PhD project focuses on the mediatization of scientists. Previously he worked as a journalist with popular science magazine *Tarkude Klubi* and *Postimees,* the most esteemed newspaper in Estonia. He received an MSc in science communication from Imperial College London. He regularly trains PhD students and researchers on science communication. He has also published about the science media coverage in Soviet Estonia and the use of open science in knowledge transfer. His other research interests include environmental communication and innovation communication.

**Shannon P. Reed** is the program manager for the NASA Space Science Education Consortium (NSSEC). She is responsible for all areas of program management, such as scheduling, staffing, and technical performance, including maintaining internal and external relationships. Ms. Reed brings to this role her success as the former deputy program manager for NSSEC during which time she coled NASA's national public engagement efforts for the 2017 total solar eclipse.

**A. Erik Stengler** belongs to the generation of astrophysicists whose career was inspired by Carl Sagan. As an astronomer, he worked in the field of observational cosmology with Professor Alexander Boksenberg using data from two space-based observatories: the International Ultraviolet Explorer and the Hubble Space Telescope. Dr. Stengler now teaches the science museum studies track at SUNY Oneonta's Cooperstown Graduate Program, situating science museums, science centers, and planetariums in the wider context of science communication. He has extensive experience in science museum education, programming, exhibition design, and outreach, having led several large publicly funded projects to take museum activities to underserved communities or unexpected places. Dr. Stengler has published in both astronomy and science communication, including articles and book chapters, and has served as editor of conference proceeding volumes.

**Jennifer Vaughn** is The Planetary Society's chief operating officer (COO). Vaughn started her career with The Planetary Society in 1997, shortly after graduating from Loyola Marymount University. Her early work at The Planetary Society was primarily with the organization's magazine, *The Planetary Report*, and the website, planetary.org. In 2004, Ms. Vaughn became the director of publications for the organization, overseeing the magazine, website, radio show, and social media channels. She took over as COO in 2011, shortly after Louis D. Friedman retired and Bill Nye took the helm as chief executive officer. As COO, she leads the organization's strategy and operations, including the fundraising and communications campaigns that led to the successful LightSail mission.

**Michelle Viotti** is a science applications and data interaction engineer at NASA's Jet Propulsion Laboratory, a division of the California Institute of Technology. She is currently a member of NASA's Mars Exploration Program Reconnaissance Team, which focuses on research studies and robotic mission formulation that prepare for the human exploration of Mars. Among other roles, Dr. Viotti was the manager for NASA's Mars Exploration Program public engagement program from 1999 to 2017. She has received NASA's Outstanding Leader Medal (2006) and Exceptional Achievement Medal (2013) for her creative and effective leadership in providing gateways to discovery at Mars for nationwide audiences with an emphasis on diversity, educational equity, and inclusion, particularly for underserved students and families. Her doctoral degree is from the University of Southern California, her master's from Johns Hopkins, and her bachelor's from Wellesley College. Along with Mars exploration, her research interests include a blend of cognitive and computer sciences for Mars data visualization, GIS mapping, human−computer interactions, and optimal brain-based comprehension and communication techniques in computer-enabled experiences that support schema development and psychosocial outcomes—including a growing societal place-based relationship with Mars and active participation in humanity's next great leap.

**C. Alex Young** is a NASA solar astrophysicist studying space weather in our solar system and beyond. He led the NASA national education and outreach activities for the August 2017 total solar eclipse. Dr. Young is the associate director for Science in the Heliophysics Science Division at NASA's Goddard Space Flight Center and the head of the NASA Space Science Education Consortium (formerly the Heliophysics Education Consortium). He is responsible for overseeing and coordinating education and engagement in the Goddard Heliophysics Science Division as well as for the Heliophysics Division at NASA Headquarters. In addition, he works with NASA's heliophysicists to promote and support their research. Dr. Young served as a senior support scientist for several NASA missions before assuming his current position.

# About the Editor

**Amy Paige Kaminski** is a catalyst and champion for expanding public engagement with science and technology. Her passion is bringing diverse viewpoints, values, and capabilities together with recognized expert knowledge to create policies, programs, and uses of science and technology that effectively serve society's many stakeholders. As NASA's program executive for prizes, challenges, and crowdsourcing and previously as senior policy advisor to NASA's chief scientist, she has worked to develop strategies to expand the space agency's use of open innovation methods in its research, technology, and exploration activities. Before joining NASA, Dr. Kaminski served as a program examiner at the White House Office of Management and Budget with oversight of NASA's space science and education program budgets. She also held positions in the Federal Aviation Administration's commercial space transportation office and at the National Space Society and is former editor of the American Astronautical Society's *Space Times* magazine. Dr. Kaminski received her BA in Earth and Planetary Sciences from Cornell University, MA in Science, Technology, and Public Policy from The George Washington University, and MS and PhD in Science and Technology Studies from Virginia Tech. Her work, *Sharing the Shuttle with America: NASA and Public Engagement after Apollo*, won the American Institute of Aeronautics and Astronautics' 2018 History Manuscript Award and will be published as a forthcoming book. She is also an author of numerous articles and book chapters on space policy and public engagement with space exploration.

# Introduction

Space science comprises several subdisciplines that seek to comprehend the origins, structure, and evolution of the universe, including the galaxies, stars, planetary systems, and phenomena within, as well as the possibility of extraterrestrial life. In one sense, it is among the most esoteric of the sciences. The field can seem obscure in that it entails an ostensibly arcane quest for knowledge by a specialized community of practitioners using sophisticated hardware, instrumentation, computing systems, calculations, and theories. Furthermore, studies of the cosmos are not overtly relevant to daily life or pressing societal needs in the way that fields like health, climate, agricultural, and energy research are.

At the same time, space science captures the imagination of people around the world like no other scientific discipline. Long before the trappings of modern science emerged, human civilizations across time and geography forged explanations for and relationships to the night sky, some developing deep cultural and religious connections to celestial objects that have carried into the present day. The Sun, Moon, planets, and stars—and visions of access to them—have featured in many forms of artistic expression throughout human history from cave paintings to novels and comic books. Today, seminal discoveries by researchers via telescopes and spacecraft frequently dominate news headlines, and celestial spectacles turn heads upward in amazement. Web traffic, social media activity, and attention paid by other forms of mass communication reveal just how much passion exists around watching a broadcasted Mars mission landing or an eclipse from one's backyard—or even around the downgrading and reclassification of a solar system body once known as a planet.

Not everyone becomes a professional space scientist, but efforts to share the awe of the heavens with the masses have been on the agenda of astronomers and proponents of space science for several centuries. The advent of the telescope and, later, photography and spectroscopy helped to revolutionize scientific understanding of the cosmos, and by the 19th century, knowledgeable individuals were inspired to relate the latest discoveries to intrigued crowds through exhibits, performances, and orreries, while enthusiasts published fictional accounts imagining flights to newly discovered worlds (Bigg & Vanhoutte, 2017). As the early 20th century began, space science began to take on a formalized, organized, and professionalized character in North America, Europe, South Africa, and elsewhere (see, for example, Beer and Lewis (1963) and Saint-Martin (2012)), with universities forming departments focused on creating cadres of individuals with training in physics and allied scientific fields (see, for example, Cameron (2010)) and observatories with increasingly capable telescopes springing up on college campuses, in cities, and on mountaintops (see, for example, MacDonald (2017)). Planetariums open to the public also began to appear. In 1923, the first modern planetarium projector debuted at Munich's *Deutsches Museum* and was followed in quick succession by ones in Rome, Moscow, Tokyo, Pittsburgh, and many other cities around the world (Bishop, 2003).

Similarly, professional scientific organizations such as the Royal Astronomical Society (n.d.), the American Astronomical Society (n.d.), and the Astronomical Society of Southern (originally South) Africa (n.d.) focused not only on supporting knowledge sharing among professional scientists through networking, new journals, and conferences but also on popularizing scientific information for public consumption. These societies plus the Société Astronomique de France (n.d.), the Astronomical Society of South Australia (n.d.), Nihon Tenmon Gakkai (the Astronomical Society of Japan) (Renshaw & Ihara, 1997), and still others also nurtured a growing interest among "amateurs" worldwide who pursued astronomy as an avocation. Meanwhile, nearly contemporaneous rocketry breakthroughs on three continents by Robert Goddard, Konstantin Tsiolkovsky, and Hermann Oberth sparked the development of rocket enthusiast clubs worldwide. A few decades later, Wernher Von Braun, the German rocket engineer who worked for the United States Army following World War II, astutely recognized that compelling stories about both the excitement and the value of space flight could help build support for making it a reality. He collaborated with *Collier's* on a series of magazine articles and with Walt Disney to make television appearance to make the case.

As global publics embraced these opportunities for learning and enjoyment, the opening of the space frontier by the United States and the Soviet Union brought the prospects for professional and public understanding of the solar system and universe to a whole new level. These and eventually other nations start research programs using upper atmospheric balloons and sounding rockets and, ultimately, Earth-orbiting space telescopes and planet-bound probes. US government leaders were keen on promoting the open exchange of information among the world's scientists as well as sharing knowledge emerging from space science research with the American and global publics (Kaminski, 2015). In 1958, the US Congress passed legislation establishing the National Aeronautics and Space Administration (NASA) to conduct aeronautical and space activities and directed the agency to ensure "the widest practicable and appropriate dissemination of information concerning its activities and the results thereof." Eager both to show the world its achievements vis-à-vis the Soviet Union and to legitimize and seek appreciation by American citizens of the value of the space program, NASA stood up a public information office that formed close relationships with the news media, invited the press to cover launches, and distributed information on the agency's milestone activities in human space flight as well as early robotic missions to the Moon, Venus, and Mars (Ezrahi, 1990; Lewenstein, 1993). Domestically, the space agency also communicated its plans and accomplishments by way of pamphlets and publications, movies, tours of NASA facilities, and traveling exhibits packed with models and scientific demonstrations. In addition, in the wake of the Soviet Union's 1957 launch of *Sputnik*, US government organizations, universities, scientific societies, textbook companies, and planetariums also began investing heavily in connecting specifically with students at the elementary, secondary, college, and postgraduate levels to strengthen science, engineering, and mathematics education, including in aerospace and space science (Wissehr et al., 2011).

In the sixdecades that have passed since space flight became reality, the legacy of foundational efforts to inform, educate, and fulfill public interest about space science has remained strong. As more nations have established observatories and launched probes into space to explore objects and phenomena in the solar system and universe beyond, space scientists, research organizations, and formal and informal educational institutions have continued to work to connect publics worldwide in creative ways to the growing number of discoveries emerging from these space-gazing machines. National governments, recognizing a correlation between domestic scientific capacity and national security as well as global economic competitiveness, have continued to support science education to ensure a robust available workforce (see, for example, Marginson et al. (2013)). With space being a motivator of STEM learning, space and science agencies have played integral roles by providing student internships and fellowships, teacher and faculty training, and research grants in the space sciences. Space science education has also been the focus of professional astronomy societies and a proliferation of nonprofit organizations dedicated to this purpose in industrialized as well as developing countries. The number of planetariums and interactive science centers, along with public attendance at these sites for informal science learning, has grown considerably (Schultz & Slater, 2020).

While approaches to public engagement with space science resemble forms and functions present at the Space Age's start, a great deal has changed in the intervening years. The world today is a much different place than it was in 1958. The machinery of space science is ever more powerful, providing volumes of data that constantly transform understanding of the universe's workings while being produced at a rate outpacing scientists' ability to process and analyze them fully. At the same time, the bulk of space science research remains subject to government funding choices and trade-offs against other societal programs in an era that has been marked by inconsistent public trust in science and its institutions. The voices and agency of ordinary people throughout the world, meanwhile, are growing. Information and communications technology advances have revolutionized people's access to knowledge and to each other, giving platforms to the previously unheard, sparking the rapid exchange of ideas, and creating new pathways for interactions between scientific experts and the general population. Diversity and support for underrepresented and underserved populations—women, people of color, indigenous populations, disabled individuals, and economically disadvantaged groups—now have a revered place in the quest to enrich the scientific and knowledge-generation enterprise in many countries and lives everywhere. Unique professional groups, including science communications, outreach, and museum specialists, have emerged to bridge gaps and improve opportunities for public interaction with science.

These conditions form the backdrop against which contemporary public engagement with science and technology must be understood and assessed. An abundance of books for popular or academic consumption related to space science exist, from college textbooks to biographies of renowned astronomers to science fiction novels to children's primers to coffee table atlases of the universe filled with vivid imagery. But short of sci-fi works addressing the sociological or ethical implications of human technological choice, the space science literature is relatively quiet when it comes to reflecting on the field's societal impact. *Space Science and Public Engagement: 21st Century Perspectives and Opportunities* reflexively places the focus on how space science as a field of inquiry and interest relates in the modern day to the billions of individuals in the world who have not made the study of the cosmos their life's work (varyingly referred to here in aggregate as "the public" or "nonprofessionals," and adjectivally as "public," though distinct publics are specified where possible). The book's chapters address relevant questions for anyone—space scientists, outreach specialists, policy-makers, educators, and others interested—to understand the breadth, drivers, scope, and impact of current modes of public engagement with space science as the third decade of the 21st century begins. In what ways do members of the public interact with professional scientists and space science research? How has the landscape of public engagement changed from earlier times? Whose participation is welcomed, why, and how? Whose needs is public engagement serving? How is public engagement important to the success of space science and to participants themselves? And what can be learned to inform designs of future public engagement opportunities?

In this book, space science researchers, program managers, public outreach specialists, educators, and thought leaders hailing from an array of government agencies, nongovernmental organizations, universities, and other backgrounds share their experiences negotiating and adapting to present-day social, political, technoscientific, and economic conditions to connect nonprofessionals with space science. They cover the gamut of space science disciplines from astronomy to planetary science to astrobiology to solar physics. While most of the contributors are based in the United States, the book includes contributions from authors in other countries as well and presents instances of public engagement with space science that scale from local to international in reach and impact. The breadth of engagement initiatives the authors explore is extensive in format and scope, from outdoor observing of celestial phenomena, to classroom and hands-on learning opportunities, to online-based experiences examining space imagery, to the development of space flight hardware, and more. Accordingly, the engaged segments of global society featured are vast and include students of all ages, policy-makers at multiple levels of government, space enthusiasts, science skeptics, underrepresented and underserved individuals, indigenous peoples, and local populations facing opportunities or impacts presented by a space science activity. Although a dozen case studies cannot tell a complete story of public engagement with space science, the highlighted initiatives provide a broad sense of the opportunities, successes, and challenges often encountered. Through their lived experiences, the authors offer valuable lessons learned and potentially adaptable ideas for future public engagement.

The chapters demonstrate that the aims of informing and educating the public and satiating general interest in space science continue to drive many public engagement initiatives led by the space science community. These efforts, however, have evolved considerably from earlier times as they seek to strengthen connections among the science, its credentialed practitioners, and broader society. In doing so, they reveal several notable trends. Accessibility and inclusion are major themes throughout the case studies, which illustrate that ever more people, both in number and in diversity, are being involved in space science activities. Sheri Klug Boonstra and Don Boonstra, for example, show that Arizona State University is reaching underrepresented populations to participate in STEM education initiatives, while Anton Binneman and Corné Davis demonstrate a commitment to partnering with native peoples and integrating their cultural views of the cosmos as part of a broad set of community and stakeholder engagement activities surrounding the construction of a ground-based telescope array in South Africa. Alex Young and Shannon Reed explain how partnerships with organizations across the United States helped NASA engage Americans from all walks of life in witnessing the first total solar eclipse viewable across the country in nearly a century. Linda Billings, reflecting on her experience working with NASA on astrobiology and planetary defense communication, and Ron Miller, focusing on efforts to popularize space science, each emphasize the criticality of acknowledging the reality of science skeptics and doubters and the need to develop strategies to engage those beyond enthusiasts. All of these and additional chapters also underscore the contributions made not just by scientists but by educators, communications and outreach specialists, and (as addressed in the chapter by Sanlyn Buxner, Michael Fitzgerald, and Rachel Freed) amateur astronomers in facilitating public connections.

Another trend the authors accentuate is the role of new and innovative applications of technologies in enabling the sharing of space science more widely while also supporting the creation of entirely new modes of engagement. Erik Stengler's chapter on modern approaches used by planetariums and science centers shows that digital projectors have revolutionized the way people visualize the space environment, while museum and outreach specialists also have developed creative mechanisms to give individuals with vision impairments ways to sense the dimensions and contours of the universe. Internet and social media capabilities have forever extended the ability to bring the digitally linked public closer to the extraordinary phenomena in the universe, as Zolt Levay shows through his experience with disseminating Hubble Space Telescope imagery. Michelle Viotti's chapter on NASA's public engagement program for Mars exploration makes a similar revelation. These technologies, combined with the sheer power of modern tools of space exploration churning out extraordinary volumes of data, are particularly significant in that they also are prompting direct public participation in the work of space science. Lucy Fortson, Viotti, and Buxner et al. highlight how online connections allow individuals not just to get a closer look at space imagery but also to help with the processing and analysis of scientific data via citizen science projects. Meanwhile, as Arko Olesk and Mart Noorma share, small satellite technology is putting the ability to conduct affordable and purposeful in-space research quite literally into the hands of students and community groups around the world, even in places like the authors' native Estonia that do not have national space programs.

Indeed, the movement toward recognizing the public's growing role in contributing to space science is a significant development in public engagement with the field that is highlighted in several chapters. As more people become directly involved in the work of space science in a variety of ways, public engagement with space science is increasingly following tendencies seen in other areas of science and technology. For the past half century, numerous social activists and scholars have demanded that nonscientists not be viewed merely as passive consumers of science whom, if infused with appropriate knowledge, would become supporters for a scientific activity or policy. The push for a move-away from this "deficit model" of public engagement has gradually been adopted by many scientific institutions worldwide in favor of rethinking the roles of citizens within democracies to decide the rightful place of science and technology within society and developing pathways to include their involvement (see, for example, Price (1967), Nelkin (197), Irwin (2001), Jasanoff (2004), Lengwiler (2008) and Pestre (2008)). Scholars have documented this "participatory turn" over the past few decades among regulatory agencies and research institutions worldwide focused on issues ranging from the use of genetically modified foods to the development of clinical trials for AIDS patients (see, for example, Callon et al. (2009), Jasanoff (2005), Epstein (1996) and Wynne (1996)). The chapters in this book reveal that nonprofessionals are being invited not only to participate in conducting research but also to support space science projects financially at their discretion (see Jennifer Vaughan and Lou Friedman's chapter on The Planetary Society's crowdfunding activities) and to engage in dialogues hosted by government managers to better understand public perspectives and values associated with space science program choices and controversial topics (see chapters by Billings as well as Binneman and Davis). Whereas many of the initiatives documented in this book reside on the end of the engagement spectrum focused on informing, educating, and even entertaining the public vis-à-vis space science, efforts to enable more people to contribute and be heard are in keeping with trends in science and technology studies and in a growing number of organizations worldwide to conceptualize public engagement as being about empowerment, cocreation of knowledge, respect for a breadth of perspectives, and mutual learning between scientists and the public (see, for example, Rowe and Frewer (2005), American Association for the Advancement of Science (n.d.)).

This book aims to paint a picture of public engagement with space science in the modern era by featuring a range of initiatives with different public engagement formats, target audiences, and focal areas. Given the practical constraint that only a limited number of chapters could be included, gaps certainly remain. The book, I believe, would be all the richer were it to include viewpoints and experiences from more parts of the world. Gathering more perspectives, and directly, from nonprofessionals who partake in—and sometimes lead—space science research, advocacy, and outreach activities about why they participate and how they regard their roles would also be illuminating. I hope to see lots of these sorts of stories appear in print before long.

In the meantime, there is plenty contained in these pages that should be instructive for space scientists, outreach specialists, and others interested in public engagement with the field and how future efforts can be enhanced. If this book leaves you with one lasting message, it should be that modern public engagement with space science is vitally important to all involved. It is invaluable to the participants inspired, empowered, heard, and fulfilled, especially where they may have once lacked access or faced exclusion, even if unintentionally. It is instrumental to the nations and societies whose people are uplifted in both spirit and material terms by opportunities to connect with the wonders of space. And it is fundamental to the success of the scientific field itself. Indeed, public engagement is no less essential for space science's advancement than telescopes, spacecraft, supercomputers, a specially educated cadre of researchers and technologists, or any of the other commonly considered trappings of the field. Even if their roles and contributions have not seemed as apparent or profound as those of career scientists, the involvement of various segments of society is undeniably fueling the field's continued viability and growth, whether through preparation of the next generation of space science professionals or by the contributions of nonprofessionals to the science as auxiliary data analysts or financiers. Connecting with general audiences can keep space projects in the public consciousness while interacting deliberately with specific groups helps space scientists and program managers as they seek support for their work, particularly when they act with heightened sensitivity to societal values, cultural norms, and public skepticism and concerns and seek to arrive at decisions diverse stakeholders will find palatable. In sum, this book shows that those who do not make space science their life's work are strategic allies and that engaging them in the field must never be an afterthought. Future public engagement efforts can take inspiration from the experiences shared here to develop creative ways to enable everyone everywhere to join in the space science enterprise. If designed thoughtfully, such initiatives will serve the needs of science, participants, and society alike as they support the democratization of science and affirm through action that space truly belongs to us all.

## Acknowledgments

I would like to thank the authors for their outstanding contributions to this book, which together create a portrait of the vibrant landscape of public engagement with space science today. Much appreciation goes to Marisa LaFleur, Ruby Smith, Kiruthika Govindaraju, and Peter Llewellyn at Elsevier for their guidance and patience as the book took shape, to Therese Griebel for supporting my interest in working on this project, and to Katie Spear for helping me do so in compliance with US government ethics rules. No US government resources were used in the creation of this work. I also wish to express my deep gratitude to a few special people who kept me going to the finish: my parents Cheryl and James Snyder, my daughter Maya, Richard Wong, Yvette Neisser, and Amy Harbison.

Lastly, I want to thank my colleagues at NASA and elsewhere who inspire me each day through their work in and commitment to public engagement with space science and other areas of science and technology. I hope this book proves insightful.

**Amy Paige Kaminski**
*Arlington, Virginia, United States*

# References

American Association for the Advancement of Science. (n.d.). *Communication toolkit: Why public engagement matters*. Retrieved December 18, 2020, from https://www.aaas.org/resources/communication-toolkit/what-public-engagement.

American Astronomical Society. (n.d.). *The origins of the AAS*. Retrieved December 18, 2020, from https://aas.org/about/origins-aas.

Astronomical Society of South Australia. (n.d.). *Welcome to the Astronomical Society of South Australia*. Retrieved December 18, 2020, from https://www.assa.org.au/.

Astronomical Society of Southern Africa. (n.d.). *About the Astronomical Society of Southern Africa*. Retrieved December 18, 2020, from https://assa.saao.ac.za/about/.

Beer, J. J., & Lewis, W. D. (1963). Aspects of the professionalization of science. *Daedalus, 92*(4), 764–784.

Bigg, C., & Vanhoutte, K. (2017). Spectacular astronomy. *Early Popular Visual Culture, 15*(2), 115–124.

Bishop, J. E. (2003). Pre-college education in the United States in the twentieth century. In A. Heck (Ed.), *Information handling astronomy — historical vistas* (pp. 207–231). Kluwer.

Callon, M., Lascoumes, P., & Barthe, Y. (2009). *Acting in an uncertain world: An essay on technical democracy*. trans. Burchell, G. MIT

Cameron, G. L. (2010). *Public skies: telescopes and the popularization of astronomy in the twentieth century* [Unpublished doctoral dissertation]. Iowa State.

Epstein, S. (1996). *Impure science: AIDS, activism, and the politics of knowledge*. University of California.

Ezrahi, Y. (1990). *The descent of Icarus: Science and the transformation of contemporary democracy. Harvard.*

Irwin, A. (2001). Constructing the scientific citizen: Science and democracy in the biosciences. *Public Understanding of Science, 10*(1), 1–18.

Jasanoff, S. (2004). Science and citizenship: A new synergy. *Science and Public Policy, 31*(2), 90–94.

Jasanoff, S. (2005). *Designs on nature: Science and democracy in Europe and the United States. Princeton.*

Kaminski, A. (2015). *Sharing the shuttle with America: NASA and public engagement after Apollo* [Unpublished doctoral dissertation]. Virginia Tech.

Lengwiler, M. (2008). Participatory approaches in science and technology: Historical origins and current practices in critical perspective. *Science, Technology, and Human Values, 33*(2), 186–200.

Lewenstein, B. V. (1993). NASA and the public understanding of space science. *Journal of the British Interplanetary Society, 46*, 251–254.

MacDonald, A. (2017). *The long space age: The economic origins of space exploration from colonial America to the cold war*. Yale.

Marginson, S., Tytler, R., Freeman, B., & Roberts, K. (2013). *STEM: Country comparisons: International comparisons of science, technology, engineering and mathematics (STEM) education. Final report*. Australian Council of Learned Academies.

Nelkin, D. (1975). The political impact of technical expertise. *Social Studies of Science, 5*(1), 35–54.

Pestre, D. (2008). Challenges for democratic management of techno science: Governance, participation, and the political today. *Science as Culture, 17*(2), 101–119.

Price, D. K. (1967). *The scientific estate. Harvard.*

Renshaw, S. L., & Ihara, S. (1997). A brief history of amateur astronomy in Japan. *Sky and Telescope, 93*(3), 104–108.

Rowe, G., & Frewer, L. J. (2005). A typology of public engagement mechanisms. *Science, Technology and Human Values, 30*(2), 251–290.

Royal Astronomical Society. (n.d.) *A brief history.* Retrieved December 18, 2020, from https://ras.ac.uk/about-the-ras/a-brief-history.

Saint-Martin, A. (2012). French astronomy in the Belle Époque: Professionalization of a scientific activity. *Sociologie du Travail, 54*(1), 53–72.

Schultz, S. K., & Slater, T. F. (2020). Who are the planetarians? A demographic survey of planetarium-based astronomy educators. *Journal of Astronomy & Earth Sciences Education, 7*(1), 25–30.

Société Astronomique de France. (n.d.). *About the Société Astronomique de France.* Retrieved December 18, 2020, from https://saf-astronomie.fr/en-french-astronomical-society/.

Wissehr, C., Concannon, J., & Barrow, L. H. (2011). Looking back at the Sputnik era and its impact on science education. *School Science and Mathematics, 111*(7), 368–375.

Wynne, B. (1996). May the sheep safely graze? A reflexive view of the expert-lay knowledge divide. In S. Lash, B. Szersynski, & J. B. Wynne (Eds.), *Risk, environment and modernity: Towards a new ecology* (pp. 44–83). Sage.

# A Space to Explore: Mars Public Engagement Strategies for a Spacefaring Society*,#

Michelle A. Viotti

Have you ever seen the Moon eclipse the Sun, watched a Mars rover land, or sent your name on a spacecraft to a solar system destination far away? Even if you did not catch them "live," chances are that you witnessed several celestial and mission events over the past two decades, replayed in ever better resolution, across more channels and devices, and in more schools and neighborhoods than ever before. Through these years, space-related wondering and wandering have accelerated with culture and technology, enabling multiple ways for people to participate directly in discovery and feel a sense of presence beyond our home world.

Our experiential opportunities have expanded from viewing, volunteering, and visualizing all the way to visiting—at least virtually for now. We may soon see the first human explorers (ad)venturing far beyond Earth and its Moon, all the way to Mars. We will begin to see through their eyes, just as we have through the cameras of our way-finding orbiters, landers, and rovers. As we approach a time when our sustained robotic presence in space joins the prospect of a continuous human presence beyond near-Earth orbit, our consciousness about our connections to places beyond Earth expands.

Yet, what does it mean to belong to a global spacefaring culture, with rapidly evolving ways of encountering not just other places but also ourselves in relation to them? How can public engagement in space exploration help create bridges between the discoveries of our age and what they mean to us? How can people participate in this great human endeavor, not just witnessing it but participating in a personally meaningful way? Understanding our "place in space" is not just about knowing more. A bigger question for public engagement in space exploration is: *How do we form relationships to places we have never been before and may never go?*

To know where we are going in public engagement, it is helpful to trace where we have been. While public engagement innovations come from many planetary

---

* The research was carried out at the Jet Propulsion Laboratory, California Institute of Technology, under a contract with the National Aeronautics and Space Administration (80NM0018D0004).
# © 2020 California Institute of Technology. Government sponsorship acknowledged.

missions, Mars exploration serves as a good narrative case for following their evolution. In 2000, the NASA organized Mars robotic missions under a single Mars Exploration Program (MEP). The prospect of a new era for Mars kickstarted thoughts at NASA about how to create an equally ambitious effort to bring the public along on this journey. In line with its restructuring, MEP made the decision to combine all of its education and public outreach (E/PO) efforts, previously conducted separately by mission and science instrument teams. This programmatic approach avoided "reinventing the E/PO wheel" for each mission, allowed continuity in programming beyond mission end dates, and enabled the development of stable, common infrastructures with long-term partners.

Helped along by a continuous string of missions that kept Mars in the public eye, making stronger public connections to Mars has had much to do with the philosophical, strategic, and organizational underpinnings of NASA's MEP public engagement efforts. As directed, the MEP Public Engagement Plan (Viotti, 2002) was one of NASA's first to lay out a multiyear roadmap with concrete, measurable near-, mid-, and long-term public engagement benchmarks flexible enough to adjust to advances in the evolving Mars mission set and concurrent technological and social trends. The term "public engagement"[1] was specifically chosen to recognize that "public outreach" was no longer one-directional and that the public could participate directly in space exploration.

A signature approach guiding the MEP Public Engagement Plan is found in its two-part vision: "to share the adventure" and "to make Mars a real place, as real as your own backyard." The first was intended to open up direct opportunities for public participation in Mars exploration; the second to create deep personal relationships to Mars by considering it not simply as "object" with scientifically discoverable characteristics, but as a relatable place, experienced firsthand in a personal way. Whether Mars public engagement activities were directly tied to the plan or were parallel, complementary NASA efforts involving media, social media, and citizen science (the latter two did not exist at the time of the plan's writing), these two themes are lenses through which a growing public connection to Mars can be viewed.

Advances in information technology inseparably intertwine with both themes. From 2000 on, we progressively saw the emergence of the Internet, high definition (HD), 4K resolution, augmented reality and virtual reality (AR/VR), immersive worlds, natural user interfaces, smart phones, social media, big data, the cloud, open government, wearable technologies, and the Internet of Things (IoT). This vast inflation of the available means of disseminating discovery has transformed both access and experience. That has enabled the NASA to evolve from delivering compelling front-row-seat viewings to more participative and immersive firsthand

---

[1] The use of this term preceded a wider adoption at NASA and was a precursor in NASA's trend toward increasing public participation in a number of areas, including citizen science and innovation challenges. Media relations were separately handled.

experiences. Mars is no longer the province of a small cadre of planetary science and engineering experts; it has become a place that is more and more familiar as discoveries regularly flood the news and social media, as citizen scientists pore over the latest data from orbit or the surface, and as rovers give us a sense of presence there.

This chapter first provides a brief historical view of our human connections to Mars. It then explores the rise of modern Mars public engagement, the organizational and philosophical underpinnings of its success, and strategies drawn from it that can guide building ever closer place-based relationships with a neighboring planet we are on the verge of exploring in person—and even settling—as early as the 2030s.

## A Brief History of Mars Public Engagement
### From Ancient to Modern Times

Mars is a special destination in the solar system. Looking toward our future, it is the only other planet on which humans may one day live and work. Looking toward our past, Mars has permeated pop culture for millennia. Metaphorically speaking, our relationship to Mars is epigenetic. It may not literally be "in our DNA" and transferrable across generations, but our connection to Mars goes far back in human history across cultures; it may have changed in expression, but it is still with us.

That is because observation is one of the most fundamental ways to sense and make sense of our world, and Mars has always called our attention. The earliest peoples noticed Mars' reddish color in the night sky and its strange looping path in the heavens—what we know today as retrograde motion. Most found it an ominous sign, associating it with war and want, famine and floods, pestilence, and other bad tidings. Through the ages, myths of Mars and its association with various gods kept it in the public consciousness throughout world cultures.

Our understandings of our connection to the Sun and the other wandering worlds in the night sky grew. Skip from the ancients to Nicholas Copernicus in 1512, whose observations of Mars' changing brightness and retrograde motion led to altering our understanding of Earth's place in the universe forever—the dawning idea that Earth and all planets revolve around the Sun. In some ways, Mars lived up to its "bad boy" reputation in bringing grief, as that idea proved far from popular (ask Galileo, strong-armed during the Roman Inquisition into recanting a heliocentric model and sentenced to house arrest for life). Yet, in this way, Mars played a crucial role in seeing ourselves differently, as part of a solar system. Nearly a century after Copernicus, in the age of the first telescopes, Galileo was the first to see Mars as a disk rather than as a twinkling dot. Astronomer Johannes Kepler began to model Mars' elliptical path around the Sun, and Christiaan Huygens and Giovanni Cassini documented the length of a martian day. Distributed through university systems and print of the time, knowledge circulated at least among astronomers and scientists of the day. Seeing through imagination as much as through hypotheses became powerful.

Knowing now this world as place, Huygens in *Cosmotheoros* (1698) imagined the "rational Creatures, Geometricians and Musicians" who might live there. For two centuries more, telescope viewers traced the dim shades of the martian landscape, prompting early scientific hypotheses about potential vegetation, water, and intelligent life. Several suggested ways to signal martian neighbors of our own existence: growing high-contrast plants in a giant right triangle (Carl Friedrich Gauss); digging giant circular ditches on Earth and setting them aflame (Joseph Johann von Littrow); and, far less friendly, focusing a giant mirror to burn geometric patterns into the martian surface (Charles Cros). Fueled by a poor translation of Angelo Secchi's and Giovanni Schiaparelli's observed *canali* ("channels") into "canals," numerous astronomers had visions of Mars as a habitable—and sometimes inhabited—place. As late as 1907, the *Wall Street Journal*, *New York Times*, and *Century* magazines all carried stories of intelligent life based on observations by Percival Lowell. In 1911, the *New York Times* even reported ongoing canal-building on Mars by master engineers (Proctor, 1911). A print relationship with Mars soon went to radio, with an on-air dramatization of H.G. Wells' *The War of the Worlds* causing panic in 1938 among listeners who believed that an alien invasion of Mars was real. Countless science fiction stories from the 1800s and 1900s, joined by movies in the 1900s and early 2000s, carried the dream of Mars forward. Undeniably, Mars public engagement has benefitted from this "pop-culture Mars" appeal through time.

Modern spaceflight has progressively sealed our relationship with this neighboring world. Venturing out from the early days of the space age to today, we have taken our human understandings of Mars from science fiction to science fact— and perhaps to a future-forward blend of the two, where imagination again fuels our quest for intelligent life, but with ourselves as the civilization on Mars. In early Mars exploration, we thought of Mars as a lunar-like surface (*Mariner 4*), seeing only craters, and no canals (*Mariner 6 and 7*). With views of volcanoes, Mars was no longer lunarlike (*Mariner 9*), revealing itself as an alien desert with signs of impacts, volcanism, ancient water, and a carbon dioxide atmosphere (*Viking*). Witnessing evidence of past water and dynamic changes on Mars (*Mars Pathfinder, Mars Global Surveyor, 2001 Mars Odyssey, the Mars Exploration Rovers Spirit and Opportunity*) and touching water ice on the surface of Mars for the first time (*Mars Phoenix Lander*), we established that Mars has had a key ingredient for life as we know it. We moved toward finding organics, the chemical building blocks of life (*Mars Science Laboratory* and its *Curiosity* rover), and began planning for a life-detecting rover (Mars 2020's *Perseverance* and its scouting helicopter *Ingenuity*), a sample return, and the first human outpost. Ultimately, with the interest of world nations, we are building toward a sustained human presence on Mars and even imagining humans as an interplanetary species, with Mars as an eventual base from which we can travel in protective "motherships" to flyby or orbit solar system destinations even farther out.

## Genesis of Modern Mars Public Engagement

After a dearth of Mars missions since the *Vikings* in the 1970s, the 1990s set the stage for a new era in Mars public engagement, with some key mission events that transformed its nature, bringing Mars to the forefront in the public mind. On August 6, 1996, a media firestorm swept across Earth when scientists reported purported signs of microbial life in a meteorite from Mars. Scientific debate would later call the findings into question, but the Allan Hills meteorite (ALH84001) in essence hit our home world a second time, inspiring a renewed enthusiasm for all things red planet. In the midst of this fever, Mars rover *Sojourner* and the *Mars Global Surveyor* orbiter arrived at Mars to reacquaint us with our planetary cousin. These two Mars missions joined a renewal in planetary exploration: *Genesis* was studying the Sun, *Cassini* was on its long journey to Saturn, *Deep Space I* was doing asteroid and comet flybys, and *Stardust* was just about to collect its second sample of interstellar dust to return to Earth.

During this time, NASA's Office of Space Science (OSS, whose functions have since been folded into the Science Mission Directorate) was working to strengthen the return on public investment in NASA science through its education and public outreach (E/PO) efforts. In late 1993, OSS had begun a formal E/PO program (Rosendahl et al., 2004) intended to involve NASA-affiliated scientists in increasing public scientific literacy (NASA, 1996). Culturally coinciding with "a thousand points of light" volunteerism (Bush, 1988), the nationwide reach of OSS's E/PO program was just getting started through large, coordinated networks of local partners, extending into the heart of American communities. An early example, the Galileo Ambassadors Program began with a single wish dropped into a suggestion box by an inspired teacher that trained members of the public could help build public awareness of the mission to Jupiter (NASA/JPL-Caltech, 2004); it grew over time into today's Solar System Ambassadors program, with more than 1000 volunteers sharing NASA planetary missions and discoveries through nationwide community events. From a more loosely organized ecosystem of E/PO providers, OSS created a more structured "support network" across the country, and it soon became obligatory for its missions to dedicate 1%–2% of mission funds (minus launch vehicle costs) to E/PO (NASA, 2002).

Plans for public outreach included the traditional (such as mission-related classroom activities, handout materials for booths at public events, public talks, and answers to mailed-in questions) but heralded the emerging. For example, Mars *Pathfinder* began a tradition, to be carried on by later missions, of engaging students in naming Mars rovers through essay contests. Toymakers joined in the excitement, with Mattel producing a *Hotwheels* rover replica, and Uncle Milton toys debuting a Mars habitat tent and related gear for kids dreaming of being astronauts on Mars. Disneyland sponsored a Mars exhibit in Tomorrowland, and Epcot later created a Mars ride. Future Mars missions would see ongoing Mattel toys, LEGO kits, Mars-themed video games, Mars movies, and more.

More than anything, though, Mars *Pathfinder* marked the beginning of a modern era of mass communications centered on providing a "front-row seat" to discovery, fostered in large part by the rise of the World Wide Web. A few years earlier, when Comet Shoemaker Levy impacted Jupiter in 1994, wide public interest in that celestial event kicked web planning into high gear (Dunbar, 2017). What sounds obvious now was not then. NASA distributed images of the comet to the press through its "Spacelink" via Telnet or modem, as well as through a very basic web page at the Jet Propulsion Laboratory (JPL) where the media could get the latest press releases, press event schedules, and images. JPL's Planetary Photojournal was in its first version; NASA's Astronomy Picture of the Day website would not launch until June 1995, and the Hubble Space Telescope would not begin a gallery of at least one image a month until 1998 (Space Telescope Science Institute, 1998). Overall, about 10,000 websites were up on the World Wide Web, and the first commercial browser had just launched (Herbert, 2014).

Even absent social media, sufficient bandwidth, mobile devices, and other systems enjoyed today, and despite annoying delays to download data, a record 45 million people visited the page over the week following the Mars *Pathfinder* landing. It was the single most popular web event at the time, beating the Olympics the year before and other big events such as the chess match between Kasparov and an IBM computer (Harmon, 1997). Anticipated traffic necessitated linking NASA's homepage and the JPL mission site to "mirror sites" around the country to manage Internet demand (see Fig. 1.1). All Mars landings since have also rivaled the largest Internet events of their times, with strategies to cope with enormous volumes of web traffic.

In one of the earliest examples of online citizen science long before it was called that, the Mars Pathfinder E/PO team asked amateur and professional astronomers to train their telescopes and cameras on Mars to help scientists and engineers monitor atmospheric conditions on Mars up to and through *Mars Pathfinder*'s landing uploading onto the *Pathfinder* website through file transfer protocol. To this day, amateur astronomers rally when needed, such as helping to observe when a comet breezed by Mars in 2015.

*Pathfinder* scientists and engineers also initiated a cultural shift in how NASA represents itself in public view. In the context of a more previously button-upped aerospace culture, some internal NASA naysayers critiqued the mission team for seeming too casual at a moment of great national interest, *Pathfinder*'s Mars landing, on July 4 (1997) no less. As members of a largely postboomer generation, they showed up as themselves—eating traditional "lucky peanuts," as had JPL mission team members during perilous spacecraft events since 1964, but this time on camera. They came to mission control on landing night wearing casual and comfortable clothes not uncommon in the dusty indoor simulated Mars surface robotics lab but certainly unthinkable in the white-shirt-and-tie Apollo days. In doing so, they unconsciously connected directly with the audience as real and relatable, paving the way for future Mars mission team members to be themselves and, in doing so, representing America (for example, a *Curiosity* rover engineer with a creative

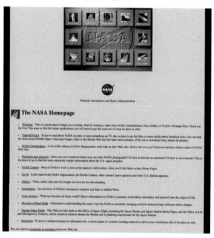

FIG. 1.1

The humble beginnings of modern public engagement: the Mars pathfinder homepage at the time of its landing (top); the NASA homepage prior to landing (middle), and the replacement NASA homepage that had to display only links to Pathfinder mission mirror sites due to unprecedented Internet demand (bottom). *NASA.*

hairstyle became the internet sensation "Mohawk Guy"). The public also had a webcam view of mission operations—a lot like the "panda cams" of today, sometimes showing a whole lot of nothing, sometimes giving a glimpse into busy scientific discussions. This way of sharing "the ordinary in the extraordinary" essentially removed a wall between rocket scientists and the rest of us. It did not hurt to feature a cute, pet-sized rover, either.

Shortly thereafter, *Mars Global Surveyor* began disseminating pictures of Mars on the World Wide Web, including one of its most compelling public engagement stories. Some of the attentive public who remembered the Mars *Viking* missions remained convinced that an image from the Cydonia region held evidence of intelligent life—the infamous "face on Mars" (see Fig. 1.2). Seen previously by lower-resolution Viking cameras, shadows on a landform suggested to some that an intelligent alien race had built a structure shaped like a face, perhaps much like the Great Sphinx, but facing up toward the skies. This small but ardent citizen group made much noise about NASA and the government purportedly hiding the truth.

Seeing familiar patterns in unclear images is a method our brains use to make sense of the world. This phenomenon, called pareidolia, includes experiences such as seeing "the man in the Moon." It is part of a rich tradition of Mars public engagement, from Lowell's telescope views of "canals" to the shapes people find in the martian terrain seen by rovers today (for example, rocks shaped like a rodent, a bird, a pyramid, or humanoids). Professionals do this, too. The *Pathfinder* team gave fun and familiar nicknames to rocks (for example, Twin Peaks, Mermaid Dune, Barnacle Bill, Yogi, Ginger, Half Dome, Cradle, Shark), as have mission teams since. It reminds us of our human need not only to make sense of patterns in the world but also to imbue them with meaning—sometimes profound, sometimes funny. The "face on Mars" event also marked a moment when the mission team shared in detail their image processing procedures to quell further doubts—that, really, it is just a hill! With regularly published, never-before-seen images of the strange and sometimes mysterious martian surface, Mars was back in the public mind.

## A Visionary Mars Public Engagement Program

From this rich cultural context, the MEP Public Engagement Plan intentionally sought to be as ambitious in providing the public with a return on its investment in Mars exploration as the two decades' worth of discoveries to come. Given the Internet suggested a "brave new world" of all things digital coming fast and furiously, developing a 20-year plan that would stand the test of time was a daunting task. The Plan was shaped in part by an advisory group made up of formal and informal educators and communications and visualization experts from leading institutions, as well as by a group of futurists from the LongNow Foundation, which envisioned responsible stewardship of Earth cultures for 10,000 years. An early online Mars E/PO public survey on NASA's Mars Exploration Program website also asked open-ended questions about why we explore Mars, people's interests,

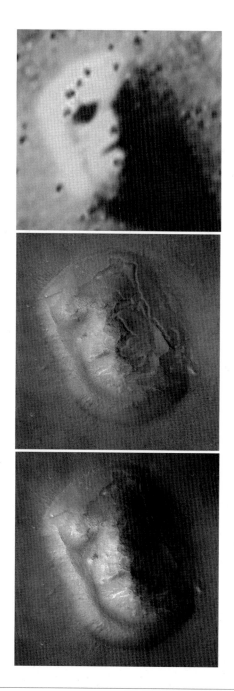

**FIG. 1.2**

Shadows on "the face on Mars" landform, from top to bottom. *Viking image (NASA/JPL-Caltech/LaRC); Mars Orbiter Camera on Mars Global Surveyor (NASA/JPL-Caltech/Malin Space Science Systems); HiRISE camera on Mars Reconnaissance Orbiter (NASA/JPL-Caltech/University of Arizona).*

perceived benefits and challenges, and what would enable them to share in the experience. A notable trend was that people already connected to Mars as a place, with one in three people citing the possibility of Mars as a chance for a second home, while many respondents offered inspirational statements about the meaning of exploration in terms of the curiosity and wonder that makes us human.

Reflections from all three pre-Plan activities (advisory group, futurist input, and public survey) to solicit innovative, audience-responsive input led to an intentional, human-centered, societal frame for public engagement that went beyond just communicating about mission discoveries and innovations, putting them in the larger context of human interest. While the Plan laid out benchmarks in each planned activity (from community to classroom, from museum to web) to ensure goal-oriented progress, success in all resulted from dedicated attention to the two-part vision: To Share the Adventure and To Make Mars a Real Place. Though these statements may sound simplistic, implementing new directions around them was anything but.

## Sharing the Adventure

When the Plan was written, mission teams enthusiastically shared information with public audiences; however, NASA/JPL communications were more cautious and controlled, particularly following the loss of the *Mars Climate Orbiter* and *Mars Polar Lander* just prior to the Plan's creation. As for making Mars familiar, the mission teams were just getting reacquainted with the red planet themselves, after two decades of absence from Mars following *Viking*. Publicly sharing the ups and downs of missions and the real human emotions that went with them was not at all an institutional practice. Such information went out through more measured press releases and still largely to the print and emerging online (presocial media) press.

Recognizing the importance of involving the public in discovery in real time given the *Pathfinder* public response, MEP's Public Engagement team (cross-disciplinary educational, communications, and multimedia experts mostly centered at JPL to enable direct connections with the missions, with connections to E/PO leads across NASA Centers and instrument teams) made an early decision to invest in HD video production in 2002, when only a few specialized channels carried that content and HD TVs were not yet ubiquitous in the consumer space. As an early adopter, MEP Public Engagement captured the engineering development of the *Spirit* and *Opportunity* rovers so that it could be preserved for later broadcasting when the technology was more available. It later enabled multiple outlets (for example, NOVA, Discovery Channel, and IMAX) to produce wide-reaching, award-winning documentaries that gave the public a window into the inner workings and real human drama of preparing to send rovers to Mars.

The intent of capturing behind-the-scenes footage was to break the stereotypical divide between "rocket scientists" and "the rest of us." While mission teams were accustomed to filming engineering tests, the camera was turned respectfully on them as people with emotions, friendships, and lives. In addition to communicating scientific and technical concepts, the raw HD video had a particular lens on the "ordinary in the extraordinary," showing scientists and engineers as regular (though highly talented and motivated!) people, experiencing relatable human emotions as they "dared mighty things" (a rallying catch phrase adopted at JPL from a speech by President Theodore Roosevelt, 1899). With prior Mars mission losses, NASA's MEP wanted to manage public expectations in the case of another bad day on Mars, reminding people that "Mars is hard." This messaging threaded through the online video series *Challenges of Getting to Mars* for all missions beginning with *2001 Mars Odyssey* and was honed for the treacherous entry, descent, and landing phase for all rover missions in *Minutes of Terror* videos. All showcased human hopes and fears in the midst of behind-the-scenes reveal of high-tech laboratories, parachute wind tunnels, desert field tests, launch towers, and other unique settings. Several TV-based "live from Mars" programs shared the drama with even wider audiences. While it took time and culture change for mission teams to feel comfortable sharing so personally, Internet affinity groups and eventually social media made sharing the personal the norm in a wider public space.

MEP also decided as a matter of policy to release raw images from Mars rovers as soon as they came back to Earth. It marked a cultural shift away from withholding almost all data releases (save for processed press-release products) for a period of time so that science teams could conduct their research and reach evidence-based conclusions. While that important research capability is still preserved, releasing raw images gave Mars exploration an immediacy and an accessibility that was unprecedented—that front-row seat everyone desired. In essence, the public could see what the rover did almost as soon as the images were returned to Earth. That gave rise to some of the first active participation in exploration. Highly motivated space fans with image-processing talent who were members of the "Unmanned Spaceflight Forum," now sponsored by The Planetary Society, often processed and disseminated Mars images and experiences from near-real-time raw images faster than the busy rover-tending mission team. While that was discomfiting at first to the science teams, today some science teams release other types of data in near-real-time as well. For instance, the public can easily find out the daily weather for *Curiosity* from an instrument onboard.

The Mars Museum Visualization Alliance was another signature MEP Public Engagement innovation that enabled museums, science centers, and planetariums to share landings and subsequent discoveries with local audiences. MEP Public Engagement had to set up a critical node outside of JPL for large downloads by museums during highly trafficked web events to ensure access at a time when the lab would be deluged and unable to support that in the precloud Internet days. Like the news media, museum partners received embargoed content so they could prepare their shows and had regular calls with mission teams so they could ask questions to support their programming. It was the local, immersive experience of the time, and you could not have audiences sitting around with nothing to watch. Before streaming capabilities, museum partners used simulated data from mission operational readiness tests as part of a critical strategy for ensuring that their real-time large-screen and dome projection systems and NASA's high-resolution data-delivery systems worked together during landing. Providing the first orchestrated "front-row seat" public viewings nationwide, this demonstration led to projections in Times Square and other popular locations worldwide that NASA frequently supports today. In recognition of this networked way of reaching local communities, NASA expanded this collaboration into today's NASA Museum Alliance, which now covers NASA-wide space exploration themes.

Providing visualizations for large-screen, immersive experiences also conceptually paved the way for more immersive technologies enjoyed today: two-way social media communications, 360 degrees videos, gigapix, and data-rich VR simulations. Early browser-based 3D gaming technologies allowed visitors to follow along rovers' paths, and a natural user-interface game allowed people to use their bodies to try to land a virtual *Curiosity* rover on Mars (see Fig. 1.3). Emerging opportunities to experience Mars firsthand are coming through wearable technologies like the hololens, which enables mission team members from around the world to collaborate—not just in a virtual meeting but together, as if walking on Mars. That ability will increasingly allow people everywhere to go from front-row-seat viewing to shared virtual visits. Room-size virtual reality experiences of the martian surface have already been tested in museums and other places of public gathering.

In addition to engagement through high-definition storytelling and immersive interactions, real public participation in mission discovery and innovation also developed over time. Essay contests gave K-12 students nationwide a role in naming all Mars rovers, conceptualizing their human meaning (*Sojourner*, *Spirit*, *Opportunity*, *Curiosity*—a tradition followed by Mars 2020's *Perseverance* rover and its helicopter companion *Ingenuity*), with wider public participation through online polling of favorite submissions. These names reflect a timelessness with currency—no stretching back to ancient Greek and Roman mythological figures or obscure acronyms but centered in the human spirit. Millions worldwide have participated by sending their own names to Mars on etched microchips carried by Mars spacecraft—not once but over many missions. While perhaps a seemingly small way to participate, it is also a large symbolic gesture of riding along on the human journey beyond our home world.

More intensive online citizen science began using *Viking* and *Mars Global Surveyor* data in the early days of NASA's "return to Mars." These efforts, which often involved crater counting and measuring by interested members of the public, were not just quantitative "make work" exercises, as they related to scientific findings about relative age-dating and understanding the history of impacts on Mars. This experience, along with the Internet's development, led MEP Public Engagement to test the public's ability to characterize publicly released raw images through an online experience cocreated with Microsoft which deliberately called on individuals to "Be A Martian"—that is, not just to process data but to identify themselves as discoverer-participants connected to place. This early experience in citizen science took place around the same time that "crowdsourcing" first entered the vernacular,

FIG. 1.3

Immersive Mars games and AR/VR experiences. *NASA/JPL-Caltech (top); NASA/JPL-Caltech/Microsoft (middle, bottom).*

but developments in image analysis quickly became more sophisticated. The *Mars Reconnaissance Orbiter* science team upped the ante with modern-day Mars citizen science: asking the public to characterize specific features in the martian terrain that computers still today cannot successfully identify on their own, directly using public suggestions on where to target its high-resolution camera based on public analysis of regional images from another onboard camera. (For missions up to *Mars Reconnaissance Orbiter*, NASA allowed supplemental E/PO funds for instrument teams to support the MEP Public Engagement Plan.) On the formal education front, one of MEP Public Engagement's longest running efforts, the Mars Student Imaging Project (see Klug Boonstra and Boonstra, Chapter 5), engaged classrooms in authentic research using the image-targeting and analysis tools of the Thermal Emission Imaging System (THEMIS) camera on *Mars Odyssey* to produce scientific findings. Through in-depth teacher learning opportunities that enabled strong student mentorship, competitively selected high-school student teams also earned the opportunity to intern with rover teams during their primary missions, participating onsite at JPL in scientific analysis tasks useful to mission scientists.

"Citizen engineering" has taken flight over time, too. Low-cost circuits, 3D printers, and a trend toward "open" everything (open data, open government, open innovation, open educational resources, and more) have inspired a "maker" culture in which programming, robotics, and home manufacturing are possible and affordable. For example, the *Rove-E* rover ("E" for education), previously built by JPL's Mars engineers for public engagement demonstrations, now has easy instructions for anyone to build replicas using affordable, off-the-shelf parts and to crowdsource design improvements or additional sensors. The maker revolution has also energized NASA's efforts to harness public participation (for example, an MEP Public Engagement partnership with Makerbot/Thingiverse for a 3D-printing Mars Base Challenge). Such novice-level public engagement efforts can be gateways to more sophisticated efforts in STEM engagement and careers (for example, NASA's 3D-Printed Habitat Challenge, which demonstrated additive manufacturing concepts for future Moon and Mars outposts).

Mars public engagement has come a long way in terms of moving from information-style outreach to experiential and participative opportunities, but the public's growing relationship to Mars has also resulted from the intentional representation of Mars as a place with robotic visitors who travel and dwell there as extensions of ourselves. Behind all Mars public engagement activities is a guiding, place-based, humanistic ethos that encourages people to connect to this strange-yet-almost-familiar environment—and to themselves in the process.

## Making Mars a Real Place

In Mars public engagement, "sharing the adventure" is often easier than "making Mars a real place." The excitement of arriving at Mars and a host of engineering firsts often capture widespread public attention, but what happens once you are there, high-energy drama disappears, discovery is slower, and the landscape is relatively bland after you have taken it in? From a color perspective, tone-on-tone Mars is sometimes a hard sell. While many aspects of a rocky world are relatable, the planet's landscape is butterscotch-rusted-tan-and-brown against a dim, beige-peach, dust-filled sky. For planetary geologists, macro- and microtextures and subtle tones reveal rock and mineral types and the processes that formed them. That provides a window into the past environmental history of Mars and clues to answering some of the biggest questions for this neighboring world: *Was Mars ever a home to microbial life? Is it today? Can it be a home for humans someday?* All of that is intrinsic for experts, part of an automated knowledge that makes their connection with Mars intimate, obvious, and personal. How can this kind of relationship be translated into an everyday consciousness, along with a sense of wonder?

Behind the Plan's vision of "making Mars a real place" was a phenomenology and a research-based, transdisciplinary body of thought that speaks to a human—environmental relationship: the meaning of place, shared and individual, that emerges through sensory, emotional, cognitive, physical, psychological, cultural, and other dimensions. It is about uniting geologic knowledge with a human experience, where the objective meets the subjective in considering the landscape, where places can be treated as "profound centers of human existence," instead of being viewed solely from the objective, scientific "view from nowhere" (Adams et al., 2001; Nagel, 1986; Tuan, 1974, 2017). In Mars public engagement, this philosophy was at the center of treating Mars as a place to go and discover, not just as a planetary object to study.

If this post-Enlightenment way of thinking seems to strike at the heart of objectivity, the point is not to diverge from scientific method and repeatable, peer-reviewed analyses and experiments but to set them in a greater context of the "Why?" beyond science-disciplinary inquiry alone. Focusing on personal and place-based meaning is not to suggest an abandonment of objective science or reignite its critique through some neo-romantic view of knowing nature through self and experience alone. It is not as simple as moving from charts and graphs to look up in silence into the "mystical moist night air," to paraphrase the listener of the learned astronomer (Whitman, 1999), or believing, as the famous movie line from *Contact* goes, "They should have sent a poet" (Bradshaw et al., 1997). It is just to note, for example, that for all of the video, media, and storytelling work around *Curiosity*'s major scientific discovery of long-sought rocks that could potentially preserve the chemical building blocks of life, the Twitterverse had almost 12 times the response (retweets and likes) to a photo of a dimly blue martian sunset, captioned with T.S. Elliot's poetic calling: "Let us go then, you and I; when the evening is spread out against the sky" (see Fig. 1.4).

Using this as an illustrative example, an ideal embedded in many Mars public engagement activities is to share an appreciation for both the scientific and the humanistic aspects of Mars exploration—to respond analytically and emotionally. In this case, the point is to engage in *scientific sense-making* (through observations, interpreting the data as evidence of potential organics) and in *scientific meaning-*

**Curiosity Rover** ✔ @MarsCuriosity · 23 Jul 2015 ⌄
Laser zaps show silica-rich Mars rocks—might preserve ancient organics.
Science afoot go.nasa.gov/1TV13wd #pewpew

💬 63    🔁 1.3K    ♡ 1.9K

 **Curiosity Rover** ✔ @MarsCuriosity · 8 May 2015 ⌄
Let us go then, you and I
When the evening is spread out against the sky

**Blue sunset on Mars** go.nasa.gov/1cxeNgq

💬 443    🔁 12K    ♡ 13K

**FIG. 1.4**

Comparison of social media reactions to Mars *Curiosity* rover tweets. *NASA/JPL-Caltech.*

*making* (connecting to the "big idea" that because Mars once had the chemical and environmental preconditions for life, past or present life on Mars could be possible). The point is also to engage in *humanistic sense-making* (virtually observing and sensing the environment from a human perspective, out of which we form personal, social, spatial, aesthetic, and emotional relationships to place and to ourselves as explorers) and in *humanistic meaning-making* (the wonder of what it would profoundly and collectively mean to us as humans to know that we are not alone, that life either has existed—or even still exists—on another world).

Many engaged professionally in Mars exploration have these integrated, internalized understandings, but the humanistic side is typically expressed more in the public engagement space. Science is an important frame for our understanding of this world: an essential part of public engagement is indeed explaining the science and encouraging science, technology, education, and mathematics (STEM) learning. Humanistic perspectives can, however, draw people toward discoveries by rousing curiosity and connection as a gateway to deeper learning (see, for example, Tuan, 2017). That is because relationships with place are at the core of our being-in-world.[2] Each place, based on its particular aesthetics, conveys meaning, and getting in touch with textures of place (Adams et al., 2001) in a way that awakens wonder can illuminate our understanding of it. In Mars public engagement, curated online galleries of "Mars as Art" and related traveling exhibits capture that, accompanying discovery-based results with aesthetic appreciation. In them, we gain a human encounter with the weird and the wonderful in a way that opens up a mesmerized curiosity: "what's that?" and "wow" (see Fig. 1.5).

Where science and art intersect can be a "province of wonder," helping us grasp the unknown (Imaginary Foundation, 2010; Shlain, 2007). For understanding Mars and for scientific literacy in general, many empirical studies recognize that this integration may in fact be essential to problem-solving, creativity, critical thinking, collaboration, innovation, and transfer (Catterall, 2002; Mishra et al., 2013; Sousa & Pilecki, 2013; Partnership for 21st Century Skills, 2007). Growing advocacy for disciplinary convergence is also on the rise (Drake & Burns, 2004; National Academy of Sciences, 2018; NSF, 2017), including in federal STEM learning strategies (CoSTEM, 2018). Recent neuroscience studies also show that emotional thought is a platform for decision-making, creativity, learning, and memory (Immordino-Yang & Damasio, 2007; Immordino-Yang et al., 2018). Together, cognition, motivation, and emotion help us process, store, recall, and build on multisensory inputs. Memory is not only key to knowing; it is a "place" that we go to as well (Özkul & Humphreys, 2015). Together, art, science, and social—emotional learning create a picture of the world that integrates the concrete and objective with the hard-to-define, but compelling and present, human experience. All of these research-based, cross-disciplinary ideas, perhaps infrequently spoken and unfamiliar

---

[2] This is an idea connected to Heideggerian thought about our ability to make sense of ourselves in the world. (Heidegger, 1962).

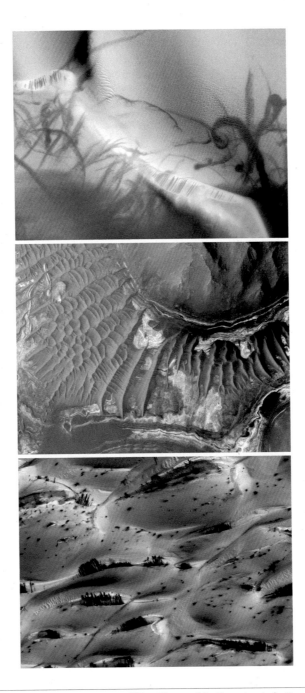

**FIG. 1.5**

The weird and the wonderful: textures of Mars from orbit. From top to bottom: dust devil tracks, dunes, and patterns from seasonally melting ice. *NASA/JPL-Caltech/Univ. of Arizona.*

in a "hard science and engineering environment," both subtly and explicitly have informed the conduct of Mars public engagement activities.

An integration of arts, sciences, psychology, and society is not unfamiliar or surprising in the world of informal education. Visit a national park, and you will find rangers engaging people with the history, geology, geography, architecture, archeology, and aesthetics of place. To make a time and place come alive, principles of interpretation (Tilden, 1957) avoid transmitting information or instructing. Instead, interpreters focus on revelation and provocation, on creating emotional relationships to places through storytelling to connect personally with visitors and their lives and to convey meanings for humanity as a whole. Storytelling helps people see the world in new ways and awakens a sense of wonder (Strauss, 1996), uniting *knowledge of* world with *being in* world. MEP Public Engagement collaborations with national parks (in particular, analogs to Mars such as the Grand Canyon, Hawai'i Volcanoes, and Death Valley) and other Mars-like places around the world sought to further these kinds of place-based, human-centric connections through video storytelling linking real field work by humans on Earth to their dreams for their robots at Mars, giving a sense of virtual presence that awakens the emotion as well as the intellect.

Martian landscapes alone, however, are still remote to human experience. It is difficult to craft a strong, meaningful, emotional, symbolic, and memory bond to a place (place attachment; see, for example, Manzo & Devine-Wright, 2014) if you have not directly experienced it. Even though we can don headphones and hear the sound of the martian wind, we will never feel it against our skin as it breezes across the rust-colored land, lifting some powder-fine dust into the thin, cold, unbreathable air. Nor would we wish to, given the lethality of that experience!

Mars robots stand in, helping to connect us to experience as much as possible. Prior to rover landings in particular, a lot of work goes into establishing their character through public naming contests and depictions of their capabilities. For example, while Mars rover *Curiosity* was being built, care was taken in artistic concepts to give the rover a kind of eponymous discovery-oriented sentience in the tilt of its "head" and in the gaze of its brown camera eyes (actually, when seen close up, not solely an artistic add but rather a reflection of a real martian surface). The intent was to begin telling the rover's story when it had not yet taken form and was in preliminary robotic pieces. It can take a while for scientific visualizers accustomed to absolute accuracy in their representations to feel comfortable taking such liberties, but when the rover reached Mars and took one of many "selfies," the factual rover ended up looking even more "sentient" in some respects than the original artwork (see Fig. 1.6).

Anthropomorphizing robots is not a cartoon trick or emotional manipulation: rovers are actually designed to give us a human perspective. Recent rovers are approximately human height (*Spirit* and *Opportunity*, the height of an average person; *Curiosity*, the average height of a professional US male basketball player). Camera "eyes" are spaced much like human eyes to enable scientists to interpret their virtual martian surroundings as if they were out on a geological field study and for engineers

**FIG. 1.6**

Postlanding Mars rover Curiosity on Mars (left) and a pre-landing artist's concept (right). *NASA/JPL-Caltech.*

to do the same in judging terrain hazards. For the people who build and care for a spacecraft or rover on its journey, it is never just a piece of metal, nor is it for a good portion of the public, either. Based on incoming heartfelt communications from the public, we all want our rovers to "wake up" and have enough energy to "get up and go," and we mourn when a rover "dies."

This idea of spacecraft as extensions of ourselves goes back to the dawn of the space age, but it is even more accessible conceptually with local drones and devices giving us unprecedented views of our own planet. Personal robots and devices are on the rise. In a way, we ourselves are becoming "human rovers," equipped with hand-held electronic sensors in our smart phones, sending pictures of ourselves-in-world to distant others who are as familiar to us as a mission team is to a rover. In participatory sensing (see, for example, Goldman et al., 2009), a citizen-science-related term carrying the personal and social immediacy of self-in-world, we offer our own place- and experience-based observations. Participatory sensing *is* about the data and being accurate; however, we should also recognize that, as people plot data about clouds, for example, they are often inspired to overlay their observations with aesthetic appreciation, and it would not be surprising if words such as billowing, towering, angry, or fleeting came to mind as they did.

Tapping into the selfie zeitgeist, rover views from waypoints and scenic overlooks give us a sense of presence. Here on Earth, we develop a spatial sense of self in time and place (Özkul, 2014; Schwartz & Halegoua, 2015) by taking smartphone pictures wherever we are and widely sharing our "identity" and experiences through location-based social media. Saved and retrievable, these images of self-in-world become instantiated memories (Frith & Kalin, 2016; Özkul & Humphreys, 2015), where sociocultural place attachment becomes more possible—not just by the individual but also by those who have registered the image of the person (or robot!) in a place, what is happening in the scene, and what meaning the image represents. When we look at a rover selfie (see Fig. 1.7), we can begin to make the same

FIG. 1.7

Place-based colors and textures of Mars, natural and robotic, and a sense "presence" in a selfie. *NASA/JPL-Caltech/MSSS.*

kind of connection and travel with it on an interplanetary field trip, feeling close to the experience.

Can these kinds of shared senses of places apply to a connection with Mars and the wider cosmos? When participating in a celebration of a celestial event or watching a Mars landing on the jumbotron in Times Square, it may indeed be possible. Something as simple as retweeting important mission-related events or images back from space are potentially "chosen markers of identity" (Schwartz & Halegoua, 2015), providing a sense of what the individual, the "friend" community, or the wider society experiences and values—in this case, exploration of a neighboring world. With emerging wearable technologies offering virtual visits, where we can meet with others in a near-real "space" and have a shared sense of (virtual) place, the possibilities grow even more.

Sociocultural "identity markers" are not always ponderous and profound. As a sign that Mars is part of our culture, part of our daily experience, and part of *us*, Mars rovers appear in TV advertisements for beer and soft drinks, industrial videos implying companies are cutting-edge, and public-service announcements inspiring kids not to shoot for the stars but for Mars. NASA social media staff are not afraid to tap into fun-loving memes of the day for social meaning and relevance (see Fig. 1.8). Meanwhile, the public online sends heartfelt "Yay!" and "Good going!"

FIG. 1.8

Loneliness meme (left); note: Curiosity only did this once, on its first landing anniversary. Meme: anonymous. Rover fist bump and sunglasses meme (right). *Left image credit: NASA/JPL-Caltech/MSSS; right image credit: NASA/JPL-Caltech.*

messages to rovers and their mission teams on landing, sustained "Happy Birthday!" messages on landing anniversaries, and sorrowful RIPs on their "deaths."

So far, robots have broken through some our Earthbound boundaries in forging a personal connection to Mars, but the coming age of human exploration beyond near-Earth orbit will expand it even more. We will eventually see ourselves on Mars through the eyes of astronauts and also through the generations who have already imagined our future there (see Fig. 1.9). For example, one of MEP Public Engagement's most sustained efforts is the Imagine Mars Project, in which learners design a human community on Mars that can not only survive but also thrive.

Integrating STEM with the arts and humanities along with empathy-based design thinking, Imagine Mars has reached tens of thousands nationwide, with special reach to underserved students living in low-income communities. Participants consider the needs of their home community (such as clean water, clean air, food, power, art, education, and fitness) before being "transported" to Mars to design solutions for the extreme martian environment; they are then "brought back" to apply their newfound problem-solving capabilities here on Earth. This experience makes Mars (and STEM learning) familiar, while encouraging literally out-of-this-world innovative, 21st-century thinking for well-being on our home world. Running now as one of the NASA Science Mission Directorate's competitively selected collaborative projects, Imagine Mars may take us all the way to seeing humans on Mars. The project's success likely lies not on students scoring higher on the next test (though great if they do), but rather on how it instills whole-person outcomes such as community-mindedness and relationships to, and solutions for, the better lives we can envision. That is, it provides a greater payoff than simply raising the next generation of STEM workers: it fosters considerations about who we want to be as a society, in all of our communities worldwide and beyond.

## The Next Phase of Public Engagement: Embracing Our Identity as a Spacefaring Society

Where do we go from here? Within a human life span, we have seen an up-close glimpse of Mars on our first planetary flybys, the first long-term orbital surveys that revealed a world so like and yet so different from our own, and the first robotic explorers on the surface. Sixteen years to the day of the Allan Hills media event, NASA saw the landing of the first rover capable of finding organics, the chemical building blocks of life. Having found organics along with signs of past and present water through multiple orbiters, rovers, and landers, we are now at the point of sending new robotic explorers capable of finding life itself on Mars, if any ever existed. This robotic exploration has led to what we enjoy today: a sustained robotic presence on and around Mars for nearly two decades, returning more data and discoveries than ever before. We have now arrived at an important moment in human history where settling on another world far beyond our own is feasible. In NASA messaging today, mission planners have gone from an earlier position that "Mars is hard" to a "#MarsReady" mentality. While our future

**FIG. 1.9**

Imagining a better future on two worlds: the Imagine Mars Project. *NASA/JPL-Caltech.*

travels do not come without risk, we know what the major challenges are, and we largely know how to (figure out how to!) solve them.

In that same span, public engagement has had similar boundary-breaking trans-formations. Twenty years from now when the first humans could be living in an outpost on Mars, our web, social media, AR/VR experiences, and other ways of connecting with people will no doubt look as quaint and old-fashioned as the first Mars webpages from 20 years ago appear to us now. We are at a similar point in time when exponential changes in computer, communications, and other technologies are on the horizon, able to transport us virtually to places wherever and whenever we are; provide us with machines able to think like (and sometimes better than) us; and supply us with wearable technologies and robotic devices that give humans the kind of "superpowers" robots enjoy.

A concrete reality of the space age is that time and technology have brought us to the point where space science is an increasingly participative social sphere. Remote sensing is not so remote anymore, and planetary exploration is not just for "rocket scientists." Our robotic explorers are truly extensions of ourselves, expanding our senses beyond our home world to places we cannot (yet) go. News and science-oriented programs bring to living rooms mysteries, novelties, and wonders on a daily basis. The word "flagship" is no longer only applied to NASA's most capable, strategic science missions, but to its most popular social media accounts as well. Astronauts on the International Space Station talk to learners in classrooms and students launch small cubesats, win engineering design challenges, create space apps, and learn to fly robotic exploration missions in summer "planetary boot camps." Symbols sent on spacecraft presage a larger sense of connection, too. Twin rovers *Spirit* and *Opportunity* each carried a sundial with the catchphrase "Two Worlds, One Sun" and the name "Mars" inscribed in world languages. People of Earth have sent their names and/or messages to the Sun, Moon, Venus, Mars, comets, asteroids, and more. Fragments of martian meteorites—blasted off the surface of Mars and into space by large impacts billions of years ago—have made their way to Earth and into the hands of Mars mission teams. In a very human gesture, teams send these pieces of Mars back to the red planet on our spacecraft (*Mars Global Surveyor* and the *Mars 2020* rover). We are the people of Earth, but we dwell consciously beyond it now as well. The next generation of explorers is poised to build and to benefit: to live as a spacefaring society.

For this coming era, a visionary prospect for public engagement is needed anew, as a human community living and working on Mars will touch all parts of our society, and it will take the talents of many people from around the world to make it possible. Providing pathways for increasingly informed and capable public participation does not happen overnight; as with technology development, it takes focus and strategy. Much like the early days of Mars public engagement, an abiding commitment to enabling public participation in discovery and exploration matters. Today's students are especially important, as they represent the "Mars generation"— those who will not only witness the first human presence on Mars but also potentially be a part of the technology-based workforce here on Earth that makes it possible.

"Sharing the adventure" and "making Mars a real place" continue to be relevant themes as future plans for Mars exploration hasten in yet another boundary-breaking era of interplanetary connections. As we truly become a spacefaring society, strategies that enrich personal participation in, and relationship to, the discoveries, inventions, and wonder to come are critical. They also increasingly depend on early integration into multimission planning, operations, and infrastructure development (for example, high data-transfer capacity between Mars and Earth). Informed by the past two decades of experience, broad recommended strategies include the following:

- **Creating a Sense of Place**: Enabling technologies like 360 degrees VR videos are increasingly common on Earth, but it takes special planning to create similar experiences on Mars. The good news is that mission teams are using advanced AR/VR technologies to place themselves "in scene." Thus, numerous high-quality imaging experiences will no doubt be available for transporting people to these virtual locations. By simply looking out across martian vistas in all directions, rover camera "eyes" can create a timeless first-person visual experience. When you want to walk around a special feature and view from all sides, say, a hill today or a human habitat tomorrow, planning in maneuvers that position rovers at various points to capture all-around views and freedom of movement within them can transform connections to Mars. Decision-making around key vistas needs more than scientific criteria and prioritization alone. Providing richer immersive content by collecting long-term cumulative data sets that enable immersive telepresence would enable the creation of experiential virtual field trips for delivery through consumer, educational, museum, gaming, and social media spaces. Especially when a few candidate human landing sites are selected and reconnaissance robots begin to patrol the area as advance scouts, tandem mission-enabling and public-engagement immersive experiences can prepare our whole society for living and working on Mars long before the first human explorers venture forth to share their real-time experience.
- **Storytelling in Scene:** Providing such visually rich immersive capabilities is critical but insufficient. Mission team members don headgear and arrive at these places with deep automated knowledge about how to move around, make sense of features and generate new meanings from them. Public engagement can bring together a number of trending practices to develop similar public understandings. Data science storytelling is emerging as a critical capability for explorers and reporters in the digital age, assisting them in creating results-driven storylines. Empathy-based design thinking entails taking a human-centric approach to collaborative solutions, starting with understanding people's needs, feelings, and motivations. Emotion is increasingly seen in neuroscience to aid in learning, memory, and meaning-making, along with place-based learning strategies and interpretative methods that connect to people's life experiences. No longer science fiction, not solely science fact, Mars public engagement can increasingly engage human-centric perspectives and participation in our collective spacefaring story.
- **Public Engagement Readiness Levels (PERLs):** In the technical world, defined technology readiness levels (TRLs) describe how well a given invention has met standards for spaceflight along a developmental continuum. Too often, public

engagement products and activities are generated in a flow but may not reflect investments in evidence-seeking tests. A similar, PERL system for public engagement technologies would encourage research-based experimentation from the cutting edge to the ubiquitous, developing shared expertise among public engagement providers and government/industry/academic partners, resulting in better quality experiences. For effectiveness in a rapidly developing, technology-enriched communications world, there is room for scholarship and experimentation in applying transdisciplinary, research-based practices. Greater impetus for this imperative exists in education but applies to all types of public engagement. For example, visual and narrative communications benefit from advances in neuroscience and educational psychology (for example, how to produce visuals in ways that make it easier for the brain to make sense and meaning and retain in memory). Effective practices vary depending on format (for example, video vs. immersive) and audiences' prior knowledge, so understanding how to select and scaffold content appropriately is key, along with designing to support orientation, spatial awareness, and user controls in interactives and immersives.

- **Public Engagement Payloads:** The MEP Public Engagement Plan recognized the importance of building payloads focused on the general public into early mission requirements, reserving mass, power, and budget for this purpose while ensuring no to low operational complexity. While Mars missions have been able to place some publicly engaging components (for example, microchips bearing the names of members of the public who registered for the opportunity) and to use instruments for public imaging and analysis, the promise of personalized, participatory exploration rises with cubesats and other small devices. As larger cargo payloads go to Mars in preparation for human exploration, student teams and the public may be able to hitch a ride for their own small-scale sensors, contributing directly to characterizing our new home on Mars. To prevent barriers to entry and to promote the development of STEM identities from foundational stages, citizen science/citizen engineering investments that develop these capabilities systemically could be critically enabling of "exploration for all."

With the prospect of human exploration, reconnaissance of Mars will expand, providing ever new opportunities for public engagement. Thousands of years from now, future peoples will remember our times as an age when humans ventured beyond our home planet, first setting foot on other worlds—when we ourselves become the "rational Creatures, Geometricians and Musicians" Huygens once envisioned living there, creating new cultural meanings and resonances. Telescope viewers may someday gaze out from Mars as early astronomers did here and watch the heavens with dark skies once more. Mars may even become a "jumping off point" from which humans venture farther out into the solar system. It is nothing less than a civilization endeavor, one that sets us on a path toward becoming an interplanetary species. Earth natives, we are on the verge of unprecedented virtual and actual voyages beyond our home world. As firsthand participation and immersive opportunities grow, future public engagement can embrace and shepherd rising ethnographic questions about how we see ourselves as members of a spacefaring society and how we extend inclusion, as exploration belongs to us all.

## References

Adams, P. C., Hoelscher, S. D., & Till, K. E. (Eds.). (2001). *Textures of place: Exploring humanist geographies*. University of Minnesota Press.

Bradshaw, J. (Executive Producer), Sagan, C., Druyan, A. (Co-Producers), & Zemeckis, R. *Contact* (1997.

Bush, George (August 18, 1988). *Speech presented at the Republican National Convention, New Orleans, LA*. Retrieved from https://www.presidency.ucsb.edu/documents/address-accepting-the-presidential-nomination-the-republican-national-convention-new.

Catterall, J. S. (2002). The arts and the transfer of learning. In R. J. Deasy (Ed.), *Critical links: Learning in the arts and student academic and social development*. Washington, DC: Arts Education Partnership.

CoSTEM. (2018). *Charting a course for success: A federal strategy for STEM education*. Washington, D.C.: National Science and Technology Council. Retrieved from https://www.whitehouse.gov/wp-content/uploads/2018/12/STEM-Education-Strategic-Plan-2018.pdf.

Drake, S. M., & Burns, R. C. (2004). *Meeting standards through integrated curriculum*. ASCD.

Dunbar, Brian (2017). *NASA: The day the internet stood still*. Retrieved from https://www.nasa.gov/specials/pathfinder20/.

Frith, J., & Kalin, J. (2016). Here, I used to be: Mobile media and practices of place-based digital memory. *Space and Culture, 19*(1), 43–45.

Goldman, J., Shilton, K., Burke, J., Estrin, D., Hansen, M., Ramanathan, N., …, West, R. (2009). *Participatory sensing: A citizen-powered approach to illuminating the patterns that shape our world*. Los Angeles: UCLA Center for Embedded Networked Sensing.

Harman, A. (July 14, 1997). *Mars landing signals defining moment in web use*. New York Times. Retrieved from https://www.nytimes.com/1997/07/14/business/mars-landing-signals-defining-moment-for-web-use.html.

Heidegger, M. (1962). *Being and time*. New York: Harper & Row.

Herbert, G. (November 2, 2014). *1994 in technology: What the internet, computers, and phones were like 20 years ago*. Retrieved from https://www.syracuse.com/news/2014/11/technology_history_internet_computers_phones_1994.html.

Imaginary Foundation. (2010). *The undivided mind*. Retrieved from http://www.fifty24sf.com/news/fifty24sf-imaginary-foundation-undivided-mind.

Immordino-Yang, M. H., Darling-Hammond, L., & Krone, C. (2018). *The brain basis for integrated social, emotional, and academic development*. National Commission on Social, Emotional, and Academic Development.

Immordino-Yang, M. H., & Damasio, A. (2007). We feel, therefore we learn: The relevance of affective and social neuroscience to education. *Mind, Brain, and Education, 1*(1), 3–10.

Manzo, L. C., & Devine-Wright, P. (2014). *Place attachment: Advances in theory, methods and applications*. Oxon: Routledge.

Mishra, P., Terry, C., Henriksen, D., & the Deep-Play Research Group. (2013). Square peg, round hole, good engineering. *TechTrends, 2*(57), 22–25.

Nagel, T. (1986). *The view from nowhere*. New York, N.Y.: Oxford University Press.

NASA. (1996). *Science in air and space: NASA science policy guide*. Washington D.C: NASA.

NASA. (2002). *Explanatory guide to the NASA office of space science education & public outreach evaluation criteria, version 1.0*. Washington D.C.: NASA.

NASA/JPL-Caltech. (2004). *NASA's diplomats*. Retrieved August 26, 2018 from https://www.nasa.gov/missions/solarsystem/f-ssa.html.

National Academies of Sciences, Engineering, and Medicine. (2018). *The integration of the humanities and arts with sciences, engineering, and medicine in higher education: Branches from the same tree*. Washington, DC: The National Academies Press. https://doi.org/10.17226/24988

NSF. (2017). *NSF's 10 big ideas: Convergence research*. https://www.nsf.gov/news/special_reports/big_ideas/index.jsp.

Özkul, D. (2014). Location as a sense of place: Everyday life, mobile, and spatial practices in urban spaces. In *Mobility and locative media* (pp. 121−136). NY: Routledge.

Özkul, D., & Humphreys, L. (2015). Record and remember: Memory and meaning-making practices through mobile media. *Mobile Media & Communication, 3*(3), 351−365.

Partnership for 21st Century Skills. (2007). *Framework for 21st century learning*. Retrieved from http://www.p21.org/documents/P21_Framework_Definitions.pdf.

Proctor, M. (1911). *Martians build two immense canals in two years: Vast engineering works accomplished in an incredibly short time by our planetary neighbors − wonders of the September sky*. Retrieved from https://www.nytimes.com/1911/08/27/archives/martians-build-two-immense-canals-in-two-years-vast-engineering.html.

Roosevelt, T. (April 10, 1899). *The strenuous life*. Chicago, IL: Speech presented at The Hamilton Club. Retrieved from https://www.bartleby.com/58/1.html.

Rosendhal, J., Sakimoto, P., Pertzborn, R., & Cooper, L. (2004). The NASA office of space science education and public outreach program. *Advances in Space Research, 34*(10), 2127−2135.

Schwartz, R., & Halegoua, G. R. (2015). The spatial self: Location-based identity performance on social media. *New Media & Society, 17*(10), 1643−1660.

Shlain, L. (2007). *Art and physics: Parallel visions in space, time, and light*. New York: Harper Perennial.

Sousa, D. A., & Pilecki, T. (2013). *From STEM to STEAM: Using brain-compatible strategies to integrate the arts*. Corwin Press.

Space Telescope Science Institute. (October 21, 1998). *Astronomers unveil colorful Hubble photo gallery*. Retrieved from http://hubblesite.org/news_release/news/1998-28.

Strauss, S. (1996). *The passionate fact: Storytelling in natural history and cultural interpretation*. Golden, CO: Fulcrum Publishing.

Tilden, F. (1957). *Interpreting our heritage*. NC: University of North Carolina Press.

Tuan, Y.-F. (1974). Space and place: Humanistic perspective. In J. Agnew, D. N. Livingston, & A. Rogers (Eds.), *Human geography. An essential anthology* (pp. 444−457). Blackwell Publishers.

Tuan, Y.-F. (2017). Humanistic geography. In *Theory and methods* (pp. 127−138). Routledge.

Viotti, M. (2002). *Mars Exploration Program (MEP) public engagement plan, D-35507*. NASA/JPL-Caltech.

Whitman, W. (1999). *Leaves of grass*. Philadelphia: David McKay. Retrieved from www.bartleby.com/142/ (Original work published 1865.).

# Science Centers and Planetariums—Bringing the Universe Within Public Reach

**A. Erik Stengler**

## Introduction

Perhaps it is no coincidence that the three forms of scientific observation that have shown the strongest capacity to inspire organized amateur activity are about looking up to the skies. All over the planet, people engage in birdwatching, cloudspotting, and stargazing—listed here in increasing distance of objects of interest from the Earth's surface, in what looks like a modern version of the celestial spheres model.

All three activities had in their origins clear practical purposes. At some point in time, technology superseded direct human observation for each of these aims. Birds can be tracked with geolocators and electronic tags, the weather is forecast based on satellite monitoring of the atmosphere, and navigation relies on GPS devices. However, direct observation has retained an irreplaceable level of fascination to many people. In terms of space sciences and astronomy, science centers and planetariums often act as hubs for the pursuit of such amateur interest and engagement, and, more recently, for complementing scientific data collection and research through citizen science initiatives.

In this chapter, I explore the landscape of public engagement in science centers and planetariums, discussing briefly some principles and ideas that have informed the past and current development of these informal education venues and that align them with trends in audience needs and expectations.

## Science Centers

### On the Path Toward Real Interactivity

In recent years, the distinction between science centers and science museums is becoming less clearly defined. For many decades, it was generally understood that *science museums* were institutions that had collections. Natural history museums, anthropology museums, and museums on the history of a specific discipline (often medicine) would count as examples thereof. At the same time, it has also been an unspoken and—in my view—unjustified assumption that the term

Space Science and Public Engagement. https://doi.org/10.1016/B978-0-12-817390-9.00011-7

*science center*—sometimes labeled explicitly as *interactive*—refers to institutions that followed the example and often the blueprints of the Exploratorium in San Francisco, with an approach in which it is not collections of objects that are the focus of exhibitions but rather *interactive* exhibits that communicate scientific concepts and principles through manipulation and "learning by doing." As science centers have only a very short history, it is only now that historians are addressing them as an object of study, and indeed, one of the clear insights that is finally finding its way to be expressed in the literature is that hands-on engagement coexisted within traditional collection-based museums, as can be seen at the Palais de la Decouverte in Paris, to name but one example. The concept of "science center" itself, with its implication of interactivity, had already appeared in the 1930s as a way to designate the New York Museum of Science and Industry (Schirrmacher, 2018).

Many science centers were opened around the world under the slogan "forbidden not to touch." However, institutions that adopted only the hands-on element of this approach soon learned that being able to manipulate objects is necessary but not sufficient: real interactivity, which is part and parcel of and leads to true experiential learning, is not only hands-on but also minds-on and hearts-on, as the late Jorge Wagensberg demonstrated in the Museum of Science of Barcelona, Spain. Minds-on represents mental engagement, while hearts-on refers to the very important and often underestimated and forgotten emotional engagement (Wagensberg, 2006). As Wagensberg liked to point out, these forms of engagement are not exclusively but best achieved through human interaction and creating opportunities for conversation—elements that need to be an integral part of professionally designed exhibitions and exhibition spaces of science centers.

In view of its success and notoriety, the Exploratorium published the famous Cookbooks (Bruman et al., 1983). These three volumes provide detailed guidance on the construction of interactive exhibits, classified by scientific discipline. In the current, expanded listing of 245 exhibits available online, the Exploratorium includes 12 exhibits directly related to astronomy and space sciences (https://www. exploratorium.edu/exhibits/subject/astronomy-and-space-sciences), but many more are indirectly related to these areas as they cover the basics and applications of optics, magnetic fields, and so forth. The original printed Cookbooks included some additional exhibits related to astronomy and space sciences, like the "telescope" or "spectra." The Cookbooks gave many new science centers a much-needed way to build their own exhibitions at a time when there were hardly any other sources of inspiration for starting up a science center. An unintended result of this initiative is that numerous projects inadvertently have ended up with a set of exhibits almost identical to the Exploratorium's exhibition or a subset thereof, for no other reason than their availability in the Cookbooks. Having a science center devoted to discovery, critical thinking, and inquiry-based learning in every major city is good, but in a connected world like that of the 21$^{st}$ century, not going much beyond the Cookbooks can limit the scope of science centers to local audiences and schools, as the visitors' perception of science centers may be that they have almost identical content everywhere, very much like public libraries.

This situation has been alleviated, in part intentionally, by approaches that differentiate science centers from each other. Some of these approaches have brought science centers to play in the league of visitor attractions, while other science centers have become visitor centers that interpret specific science-related venues. I will discuss both below. In both cases, astronomy and space sciences venues have led the way in the sector.

## Science Centers as Visitor Attractions

Whether a science center presents itself as part of the museum circuit or as a visitor attraction may seem a matter of semantics, but it has profound implications in the way it is managed and funded, and consequently for its very sustainability. It probably was a question of semantics when science centers in unison fled from the "museum" label and instead used all sorts of other names—house, hall, city, park, space, palace, factory, world, and others—but were de facto mainly designed and run as museums and were included as such in visitor guides and similar listings. But truly becoming a visitor attraction means having a relationship with a much wider range of visitor types, including tourism and leisure audiences who visit with a different mind-set than traditional museum visitors or school groups. Those who approach a science center as a visitor attraction are not necessarily so interested in educational and cultural outcomes or a continuous relationship that could lead to community-building around the science center but rather in one-off entertainment values. For example, a tourist will naturally be more interested in learning about unique features tied to the local context of the country or city they are visiting than to learn about science during that visit. As a consequence, serving as a visitor attraction also involves a different set of stakeholders and funding streams, such as local and regional councils through their tourism and development departments; industries that are part of the history of the relevant technoscientific sectors and the local socioeconomic fabric; and government agencies, such as research councils and, in the case of space sciences, national space agencies.

Science centers operating as visitor attractions face a competitive playing field, more so than those centers acting in a purely educational context, but they can work quite well. To do so, the venues need to focus on a theme, and we have examples of success stories in the space sciences, such as the Cité de l'Espace in Toulouse, France, including its IMAX theater and a planetarium (https://en.cite-espace.com/). The Cité is presented as a theme park with connections to the renowned local aerospace industry and the French space program, which make it unique and attractive to visitors who want to know more about the region and its role in space exploration. The same happens, for example, with the Cosmodome in Laval, Canada, near Montreal (https://www.cosmodome.org/en/). Both institutions market themselves as tourist attractions while still maintaining a strong educational offering that is relevant and attractive to educators in a large area. They, and many others like them, are following the trend of the economy that situates current offerings in the "experience economy" (Pine & Gilmore, 1999), according to which visitors are no longer interested in paying for a "commodity" (such as scientific information that nowadays can be found anywhere, mainly online) or a "standardized product" (such as a collection of exhibits from the Exploratorium's Cookbook), or even a "customized service" (such as a particular display of curated exhibits). They are instead seeking "unique experiences" that they would not find anywhere else and ideally not even in a repeat visit to the same venue. Coffee shops, car sellers, and travel agencies are all marketing experiences beyond coffee, vehicles, and trips, respectively, and science centers competing in the arena of visitor attractions also need visitors to exit with a memorable experience. These ideas from Pine and Gilmore were further developed and applied to the science center sector by Asger Hoeg at the Danish science center, the Experimentarium (Hoeg, 2012, 2017). At this science center, as in many others, we can see this trend materialized in exhibitions that still incorporate the "traditional" hands-on exhibits but place them in "narrative clusters" that tie them to a local context, for example, telling stories about international trade, sports, or leisure at the beach that are related and unique to a coastal city like Copenhagen.

In this framework, science centers and museums can become important collectors and preservers of unique and valuable elements that link local history, economy, industry, and scientific relevance together with people's lives. These elements can be not only objects but also images, ideas, testimonies, and oral histories. For example, the Smithsonian National Air and Space Museum keeps an invaluable collection of oral histories relating to the history of space, covering developments in science, technology, management, and politics in the United States (https://airandspace.si.edu/research/projects/oral-histories/oralhistory.cfm). Oral histories can also be collected from visitors for or even through exhibitions, as an exemplary way to allow for participation and even co-creation by the public. At the Museo de la Ciencia y el Cosmos in Tenerife, Canary Islands, Spain (https://www.museosdetenerife.org/mcc-museo-de-la-ciencia-y-el-cosmos), where I worked for a decade, we learned this with the occasion of the 50th anniversary of the solar eclipse of October 2, 1959, the last one visible from the archipelago. It all started with a program to take the topic of eclipses, as well as other activities, to the often-forgotten audience of senior citizens. We began explaining the mechanics of eclipses with great enthusiasm, but we did not get very far: everyone in the audience was more interested in telling us where they were and what they observed during the eclipse. We got the message and soon turned the program into a temporary exhibition made with their testimonies (https://www.museosdetenerife.org/mcc-museo-de-la-ciencia-y-el-cosmos/evento/1630) (Fig. 2.1).

## Science Centers as Visitor Centers

Another format that allows science centers to be unique and different from others is that of a visitor center for an area where science is happening. I am intentionally vague when I say "science is happening" because I want to include human activity as well as natural phenomena that are intrinsically relevant for science, such as volcanic areas or nature reserves. This visitor center approach also brings science

FIG. 2.1

The author introducing the topic of eclipses to senior citizens in Tenerife, Spain. *Erik Stengler.*

centers to the touristic playing field, as they offer unique experiences that cannot be found anywhere else but perhaps with a more contained, cultural flavor and less focus on entertainment.

For example, the popularity of space travel and exploration has made NASA's visitor centers tied to its flight and research facilities a very important element of the landscape of engagement with space sciences (http://www.visitnasa.com/). In the case of astronomical observatories, the desire of the public to approach and know more about real things that are happening and being done under those huge domes has led to the creation of visitor centers which often include science-center-style interactive exhibits that fulfill this interest without having to interfere with the scientists' work. The Astrolab (https://www.astrolab.qc.ca/en/) at the National Park of Mont Mégantic, Canada, serves as visitor center for the Mont Mégantic Observatory. The Astrolab is very near the observatory, so it will mostly cater to those who make their way up to the mountain top.

But there are also visitor centers that are not next to the observatories' remote sites. A creative example in this regard is a visitor center for the French observatory of Pic du Midi. The observatory has a visitor center located at the nearest highway rest area (https://www.abelard.org/france/motorway-aires16_pic_midi.php), where tourists can stretch their legs from long drives through France while learning about the research that is taking place at the top of the Pyrenees nearby. This certainly is a way to reach out to unwitting audiences. In similar fashion, the Museo de la Ciencia y el Cosmos in Tenerife was not officially conceived as such but doubles as an entry-level visitor center that complements and helps keep at a manageable size the occasional and restricted open days, special tours, and other programs run by the Institute of Astrophysics of the Canary Islands at its mountain-top observatories on the islands of Tenerife and La Palma. Sharing observatory work from an even greater distance, the European Southern Observatory (ESO), which runs telescopes at three sites in northern Chile, has recently opened the ESO Supernova Planetarium and Visitor Center near the organization's headquarters in Garching, Germany, near Munich (https://supernova.eso.org/).

## Planetariums
### Creating Communities, Inspiring Young People

Long before planetariums became a standard element of science museums and science centers, they were large and expensive facilities whose development was almost exclusively in the hands of the Carl Zeiss company in Jena, Germany, since Zeiss invented the projection planetarium in the 1920s. Many such planetariums, either as stand-alone institutions or as part of a major science museum, were for decades the reference point where young and adult amateur astronomers would meet regularly, find out about the latest news and discoveries from space exploration programs, and have direct access to the experts. The visit to the planetarium with the school or for a public event was a major highlight of the year, or perhaps a few times a year if the distance would allow it.

Many astronomy clubs and societies originated in and around planetariums and, in a perfect symbiosis, drew regular audiences to the planetarium from their membership, while first-time visitors to the shows would have the opportunity to pursue a new hobby by joining the clubs and societies. My childhood memories bring me back to such scenarios, which I associate with planetariums in Germany. During holidays, my father would take me to planetariums like the planetarium in the city of Bochum (https://www.planetarium-bochum.de/de_DE/home). We would be one-off visitors there, but I could see how the majority of people attending the show would know each other, talk to the presenting astronomer afterwards, and stay in the facilities for a while as they informally conversed about experiences, observation plans for the week, and the telescopes they used. This has changed, with many planetariums using more formal "friends of the planetarium" membership schemes that look like fundraising initiatives. The planetarium in Vienna, Austria, however, is an example of those that maintain some of that sense of community, by partnering with local community colleges and the city's two historical observatories, which are now used for education and outreach, to offer joint programming to students and astronomy enthusiasts (https://www.vhs.at/de/e/planetarium). This allows for a community to grow organically out of these institutions, with an enriching generational mix made of the astronomers who run the planetariums, retired scholars volunteering their expertise, adult amateurs, students and young kids all being able to share their passion and learn from each other. Is this not the dream of any modern public engagement initiative? Is this not the incarnation of a true "dialogue" between science and society so often discussed and demanded by science communication practitioners and science and technology studies scholars?

Planetariums are also a perfect environment to get young people interested in pursuing space sciences as a career. This is exactly how it happened to me: after a visit to the local planetarium at the nautical school in the coastal town of Pasajes, Spain (https://turismopasaia.com/en/planetarium/, at the time not open to the public and very difficult to access except through special arrangements), I was told about the regional astronomy group of the Scientific Society Aranzadi, in San Sebastián, Spain (http://www.aranzadi.eus/category/astronomia), which I joined and attended every week until I left the city to study physics and astronomy in Germany. Interestingly, I would return 12 years later to be part of the team that created and ran the science center now known as Eureka (https://www.eurekamuseoa.eus/en/home, then Kutxaespacio de la Ciencia), right next to where the astronomical society used to meet.

At a time when there were no smartphones, no Internet, no social media, and no resources one click away, planetariums were also one of the few places to look for specialized resources and materials, like books or slides. I remember very well treasuring as a child my collection of slides from the Adler Planetarium (https://www.adlerplanetarium.org/), purchased at another planetarium I had visited. How often and how long I looked at those slides projected on the wall of the living room. I used to let my imagination go wild about what other marvels of the universe would be revealed in the next available set, which I would order and await by postal mail. Progress comes at a price, and that sense of specialness I associated with my collection of slides from the Adler Planetarium is something current kids will not experience since there is a colossal wealth of images freely available online that they will skim over on their screens in very short time spans and never physically hold in their hands (Fig. 2.2).

**FIG. 2.2**

The Adler Planetarium in Chicago, Illinois. *Adler Planetarium.*

However, I would certainly not fall into the temptation of thinking that all times past were better and that somehow the digital era and the Internet revolution have spoiled an ideal world. Current planetariums are still the place to go for the present-day generation of space science enthusiasts: just as planetariums were "curating" growing collections of images and making them widely available through shows and audiovisual materials a few decades ago, it is precisely the fast-growing and almost unsurmountable amount of information available from more and more sources that makes such "curation" very necessary again, albeit for different reasons. While in the past planetariums tried to counter the scarcity of resources and materials available to the public, nowadays they are needed to counter the excess of decontextualized information and turn it into knowledge. Planetariums are the place where that wealth of information is presented in a coherent manner with historical context and in stories that make it relevant.

Planetariums have been exemplary for all areas of science and technology in the way that they opened up the famous "ivory towers" where scientists were said to hide from society. Most of the major historical planetariums that have always been part of this ecosystem of successful collaborations among scholars, amateurs, and interested members of the public are still in operation and continue playing that role, albeit less so in the form of cozy evenings between friends and regulars in badly heated facilities. Larger crowds are (fortunately) now attracted to more widely marketed shows and events, and the offerings match this new reality, including measures for comfort, health and safety, and occupancy limits. Around the globe, the passage of comets and meteor showers, the occurrence of eclipses, and observing opportunities on clear evenings continue being turned into successful public events complemented by shows under the planetarium dome.

## The Transition to Digital Theaters

To further understand the current landscape, we need to place a new generation of planetariums in the wider context of science centers. Virtually all science centers that have been created since the 1980s have felt the need to include a planetarium as part of their facilities. These planetariums have not developed around a community of astronomy scholars like those mentioned earlier but as part of business plans of institutions that need to be sustainable. As a consequence, the most economically viable option is often taken: installing a standard suite of equipment that allows the purchase of ready-made shows from one or more of the various production companies that offer regular new releases and subscriptions. This avoids the need for expensive production staff and equipment and was mainly made possible by the arrival of digital technologies. Such technologies first complemented and later fully replaced optomechanical projection systems. Although at first there was a confusing variety of digital systems that followed different concepts and often reached technological dead ends, gradually all makers have converged toward similar and compatible systems that allow easy distribution and sharing of preproduced programs. Different brands no longer compete so much in the hardware and projection systems but in the creation and distribution of such shows.

A growing number of voices in the planetarium sector have raised concerns about the move toward these "prepackaged" programs. However, this is not a new contention brought about by the availability of digital media: already when there were no digital means to "package" a show and it was done manually and mechanically with the help from increasingly complex automation systems, the debate was already on the table on whether shows should be live or prerecorded (Ratcliffe, 1996). What digital technologies have added to this debate is the question of content: once the optomechanical star and planet projector is complemented and ultimately replaced by digital full-dome projection systems, many see no reason to limit shows to astronomy, and "films" on any topic and even fiction productions are easily adapted to such an attractive immersive environment. This has been reflected by the gradual change of designation for these venues: the word "planetarium" has often given way to "digital theater." Also, the layout reflected this transformation: more and more planetariums have changed their seating arrangements from reclining seats in a circular configuration on a horizontal floor to fixed seats on a highly tilted slope, all facing the direction of a preferred area for the action to happen on the screen. The way they are operated differs very little from a regular cinema.

Even if the focus on astronomy is maintained, when a planetarium becomes a domed cinema screen where a film is shown by pressing "play," there is no need for an astronomer to be present, and when there is no astronomer, audiences will have no reason or opportunity to interact with an expert. In the absence of an astronomer to act as a catalyst for conversations, the visitors will not stay behind discussing among themselves either (Keller, 2012). Opportunities are therefore lost to promote conversations around the content presented, which is unfortunate given that conversations among visitors are one of the most valued outcomes of informal education venues, both by science communicators and by visitors themselves (see, for example, Leinhardt et al., 2002).

One way some venues have addressed this problem is to partner with the local amateur astronomy society. In the absence of a research astronomer and any links to a research institution, the role of such a group can become more prominent and important. This was taken into consideration from the outset in the aforementioned science center Eureka! in San Sebastián, Spain, where not only a planetarium but also a public observatory was included in its design. This allowed for the development of a collaboration with the well-established local astronomy society, with whom the science center undertook many projects over the years since opening in 2001. During the time I worked at this science center, joint expeditions were organized to observe and broadcast a solar eclipse (Castrillo et al., 2001) and the northern lights (Minguez et al., 2004). In this way, the science center immediately had a head start in the form of an already-constituted community of engaged users of the astronomy-related installations, and the astronomy society members saw their long-awaited dream come true of having such facilities at their disposal as a much-needed complementary public outreach platform (Fig. 2.3).

A bigger problem with this digital-theater approach is that this road has led some planetariums to compete in the field of the IMAX-dome theaters. IMAX theaters

**FIG. 2.3**

The science center Eureka! in San Sebastián, Spain, with its astronomical observatory (tower on the right) and planetarium (left). *Eureka!*.

began as a large format development of regular cinema theaters, projecting from a variant of 70-mm film stock. Ironically, the domed version of IMAX, which was initially called OMNIMAX and later IMAX-dome, was developed in the 1960s at the initiative of the San Diego Hall of Science (now the Fleet Science Center, https://www.rhfleet.org/) to project IMAX films onto their planetarium dome. Beginning in 2008, IMAX productions underwent a transition to digital, too, and the end result is that the digital IMAX-dome systems differ very little from digital planetarium systems. It is an interesting case of convergent (technological) evolution. But IMAX-domes work with production costs and running budgets that are orders of magnitude larger than those of science centers and planetariums. While many digital-theater planetariums have accepted the challenge and have launched themselves into this new playing field, many others are turning back to astronomy— or at least science—and direct contact with the public as their unique selling points. For this to happen successfully, the space science, astronomy, and science communication communities need to be there to work hand in hand with the professionalized production teams.

However, despite all issues and caveats, the ease and standardization of formats in prepackaged shows has also paved the way to new opportunities for planetariums. For example, the growing requirements from research funders to include public engagement in each project lead institutions like the Space Telescope Science Institute and the European Southern Observatory to distribute, free of charge and often in collaboration with private production companies, complete shows or easily adaptable sequences that are especially suited for these planetariums with standard digital equipment. Along these lines, planetarium makers also offer updates to vast online databases that can be incorporated into the programming and projection systems.

These databases are compiled from real astronomical data provided by research organizations, and the systems can show the newest images from space exploration and map them onto computational 3D models of astronomical bodies that audiences can then visit in virtual tours of the universe.

## College Planetariums: More than a Teaching Resource

There is another, less-known type of planetarium. This group comprises those planetariums that were built with the primary purpose of teaching, following the efforts in the 1950s by Armand Spitz, who developed a class of projectors that were affordable for educational institutions as well as smaller museums. They are often found in buildings originally dedicated to nautical and aviation studies, where they served the main purpose of teaching astronomical navigation and also supported physics and astronomy departments in universities and colleges of all sizes. The first of these, the Morehead Planetarium (https://moreheadplanetarium.org/), was pivotal for the training in astronomical navigation of the first NASA astronauts, including those from the Apollo missions (Clinard, 2019). There were many to follow. It would be impossible to compile a reasonably comprehensive list of all college and university planetariums in the world, but many of them are sticking their heads out of oblivion by joining in the provision of public-facing sessions and events.

While major public planetariums have evolved to offer automated and prepackaged shows and large-scale events in which access to research astronomers has become rare, it is in the smaller college-based planetariums where one will still find the small but engaged communities that used to originate around that first wave of public planetariums. Students, local amateurs, and general audiences attend the few public events that are offered, more as clubs than as visitor attractions. Their shows are normally very informal, with lots of dialogue during and after the sessions. They usually have an astronomy professor and graduate students actively involved and therefore provide that exceptional environment where interested visitors have direct access to the experts. Research has shown that having highly motivated, research-active astronomers involved in planetariums is an important asset just as has been the norm in other institutions like anthropology museums, where there are anthropologists on staff (Knappenberger & Duncan, 2000). As this degree of access to experts is not possible for most large public planetariums, college planetariums can focus on this as a unique selling point to create a completely different visitor experience.

There are interesting developments driven by technology in this category of planetariums, too. The availability of cutting-edge visualization technologies that are keeping planetariums of all kinds busy updating and adapting their venues also offers valuable tools for research. At the State University of New York (SUNY) Oneonta, for example, Dr. Joshua Nollenberg not only develops new planetarium shows for the public and to teach astronomy courses but also, along with his students, uses the powerful digital system for research, such as the search for a predicted ninth planet in the solar system. He does this by visualizing current research results and examining database information on the orbital parameters of Kuiper Belt objects to identify the areas of the sky in which such a planet could most likely be observed (Louden et al., 2016). This work optimizes the use of telescope observing time and helps plan new observation runs. Watching a show or attending a class in a planetarium that is also being used for current, ongoing research certainly gives audiences the opportunity to be "right there" where science is happening, and therefore offers a different, genuine, and unique experience (Fig. 2.4).

## On the Road

In addition to all the types of planetariums discussed so far, there is, as if launched from a mothership, a huge fleet of portable planetariums and other mobile resources that bring space sciences and astronomy to all corners of the globe. Just like their larger counterparts, portable planetariums have evolved along with developments in technology. What began as projections through perforated surfaces around a light bulb to simulate stars and celestial objects in their respective positions in the night sky is being complemented and often replaced by ever smaller and more affordable all-digital, full-dome projectors. What these traveling inflatable domes have maintained throughout, however, is the closeness and direct access to the presenter-projectionist. Being shoulder-to-shoulder with the presenter and sitting on the floor after crawling into the dome and having left your shoes outside certainly breaks down any expert-audience barriers and makes the planetarium show as much of a shared experience as it possibly can be.

**FIG. 2.4**

Dr Joshua Nollenberg presenting to students at the SUNY Oneonta Planetarium. *SUNY Oneonta.*

Science centers or independent science communicators regularly take portable planetariums on tour to schools, public venues, fairs, and events, bringing these informal science education opportunities to the most remote and unexpected locations. These destinations include hard-to-access territories such as small islands, prisons, holiday resorts, hospitals, nursing homes, and a long list of often-forgotten places. Along with portable planetariums, these tours also bring astronomy activities such as public observation events to these locations. For example, while I worked at the Museo de la Ciencia y el Cosmos in Tenerife, we obtained several public funding streams to bring astronomy and space sciences to the most remote corners of the Canary Islands archipelago. With a science van called "Cosmoneta" (Fig. 2.5), we toured a portable planetarium, a musical concert on the solar system called "Solar System 8.1," and a puppet theater about the Moon in the night sky.

The continued popularity of traveling planetariums and astronomy activities is a constant reminder that there are many audiences out there to be reached and that are craving for more—audiences that are as unexplored as the universe itself. From among the numerous portable planetariums in existence, I would like to mention one example that is particularly remarkable for two important reasons: one being content and the other being style. The Explorer Dome (http://www.explorerdome.co.uk/), based in Bristol, United Kingdom, breaks down not only the expert-audience barrier as mentioned before but also disciplinary boundaries (Fig. 2.6).

The Explorer Dome's sessions explore astronomy along with other related areas that take advantage in very creative ways of the special environment of the inflatable dome. They leverage the acoustic properties of the semisphere to experiment with sound and the darkened space to play with light and colors while inviting active audience participation. They combine all these elements to create interactive shows about matters that affect the Earth as a planet, such as deforestation and climate change, and touch upon topics such as biodiversity and ecosystems. The Explorer

FIG. 2.5

The author with the science van "Cosmoneta." *Sarah Callan.*

**(A)**                    **(B)**

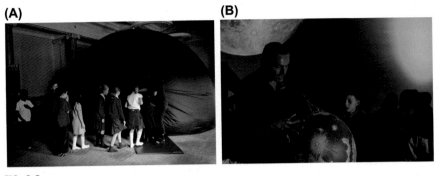

FIG. 2.6

(A) Entering the Explorer Dome, Bristol, United Kingdom. (B) Activity in the Explorer Dome. *Explorer Dome.*

Dome also treats the views of audience members with great respect. In my career, I have come across a number of scientists that get involved in public engagement with the primary aim of promoting critical thinking and debunking pseudoscience. I share these aims, but I cannot share the sometimes-aggressive approaches toward members of the public who may hold pseudoscientific beliefs. If a planetarium show or other public talk includes derogatory comments or jokes about astrology and someone in the audience believes in horoscopes, then that individual may feel alienated and be unwilling to be open to the arguments in favor of critical thinking that may follow. But most importantly, every individual deserves to be respected as a human being, and I have always been profoundly impressed and glad to see how shows at the Explorer Dome treat these issues in a tactful manner, while at the same time being firmly grounded in rational thinking and science.

## Transforming Lives
### ...Through Science Capital

I have discussed that many planetariums and science centers have landed in the highly competitive sector of visitor attractions, in which planetariums compete with IMAX theaters and cinemas, and some science centers and museums become and compete with theme parks. For many, this has been a positive move toward their own sustainability, but for many other planetariums and science centers and museums, the leisure industry market is beyond what they are able to reach or even want to enter. They strive to find uniqueness and provide memorable experiences in other ways.

One option is to focus on their educational and cultural value and fully develop their role as venues for informal learning. After all, school visits still are, for all types of science centers, museums, and planetariums, a substantial fraction of their visitor numbers and, in the majority of cases, of revenue. At the same time, the Exploratorium's original focus on experiential learning has never been more in demand than now. Educators who aspire to make their courses ever more inclusive of different learning needs and styles rely heavily on constructivist approaches.

However, classrooms are becoming more and more like informal educational settings themselves (Falk & Dierking, 2010, p. 492), and inquiry-based learning has taken over formal education. It seems that what used to be a unique selling point (in terms of education) of informal learning venues has been widely adopted in formal education. In view of this development, what added value can planetariums and science centers offer for teachers and their students?

For a long time, the answer to this question has been found at the level of motivation: having a good day at the science museum, it has been thought, will make students more likely to be interested in science at school and in their future. However, "having a good day" is, especially if it is the students who are asked in post-visit evaluations, a synonym of "having fun," and it is becoming increasingly frequent that teachers feel pressured to make their classes "fun" as well. So in this regard, too, science museums can cease to be special.

As a consequence, there have been many attempts to find a renewed uniqueness of purpose for informal science education venues within current trends in technology, science communication, and science education, and there have been as many dead ends in terms of answering the question of what science centers are for. But a new idea is emerging and taking hold in the sector that seems to be a breath of fresh air. This idea is *science capital*. It has been researched at King's College London and University College London (Archer et al., 2015) and applied to science museums and centers by the Science Museum Group (https://group.sciencemuseum.org.uk/our-work/our-approach-and-science-capital/, initially composed of various museums in the United Kingdom and now being joined by colleagues in the United States as well. Drawing on the idea on cultural capital (Bourdieu, 1986; Bourdieu & Passeron, 1977), the Science Museum Group's research explored the factors that most influenced children's intentions to take up science in their studies and found the answer in the combination of science-related knowledge, attitudes, experiences, and resources acquired through life. This is what defines science capital, and it happens through exposure to a series of elements such as family science skills, knowledge, and qualifications; personal knowledge of people in science-related jobs and roles; in-school and out-of-school activities; consumption of science-related media; conversations with others about science; science literacy enhancing actions; knowledge of the transferability of science skills; and acquired science-related attitudes and values (Science Museum Group, 2018). Science museums and planetariums can have an important role to play by providing access to unique informal learning experiences as part of a wider framework that includes several other players, such as family, friends, the media, and schools. None of them will be able single-handedly to build science capital for a person but together can make a difference and transform the lives for many citizens of the next generation.

Being able to aim for such a transformation lays out for science centers and planetariums a perfect way to align what they do with the next step of what audiences demand: following the experience economy mentioned earlier, current trends highlight the transformation economy. Audiences now seek and expect personally *transformative* experiences (Pine & Gilmore, 1999). The proliferation of self-help books and online media is a clear sign of these times. In developing science capital, the sector of informal education (including science museums, science centers, and planetariums) offers visitors the opportunity for far-reaching and profound personal transformation, and placing within reach, at last, the Holy Grail of getting more young people interested in pursuing science as a career, which is the ultimate transformation.

Finally, science centers and the science capital they foster may go a long way toward transforming society by fulfilling an equalizing role and bringing diversity and inclusion to science education. It is well known in the science education sector that, as Falk and Dierking (2010) pointed out, up to 95% of science learning takes place outside of schools. Where the formal education sector has so far been unable to break the barriers of class and social exclusion, perhaps informal education can make the difference.

## ...Through Citizen Science

As I expressed at the beginning of this chapter, I hold the opinion that the difference between science museums and science centers is becoming blurred and that this is a good thing. However, at first sight, unless they are part of a research institution, science centers and planetariums cannot aspire to house research activity in the way traditional collection-based museums do. In recent years, many attempts have been made to introduce research in science centers, often making them laboratories for social sciences or establishing links with science communication research centers. This has led to successful collaborations and valuable research, but it does not fulfill the desire of many visitors for closeness to research related to the content of the exhibitions.

An approach that is taking hold and at the same time offers the public a truly transformative experience is citizen science. By becoming hubs of citizen science projects, science centers and planetariums not only gain the ability to house research activity but also give their visitors unique opportunities to engage in real research projects, going beyond "learning by doing" and becoming direct participants in real research.

To realize this potential, it would be a giant step if new science centers and planetariums were to include facilities where researchers could train citizen scientists in the research they would take part in and were to complement this with exhibitions about the relevant topics. The Adler Planetarium is leading the way in this endeavor, going even further by embedding citizen science in its exhibitions with ingenious interfaces through which visitors can, for example, try out an adapted version of the Zooniverse online citizen science project platform (https://www.zooniverse. org/) and sign up for the full version if they desire to pursue it further from their own homes. Zooniverse has become renowned as a hub where researchers can post projects where citizen participation is essential.

## ...Through Universal Design

Another example of informal learning making a difference is the team of astronomers led by Dr. Amelia Ortiz Gil from the University of Valencia, Spain, who developed a kit of activities on astronomy for children with visual impairments. The project is known as "A Touch of the Universe" (http://astrokit.uv.es). The kit includes a planetarium program called "The Sky in Your Hands" that can be followed in a tactile manner by blind members of the audience on a 3D-printed half-sphere. With funding and support from the Office of Astronomy for Development (a joint project of the International Astronomical Union and the South African government), the kit has been distributed since 2013 among educators and teachers in underdeveloped regions of the Americas, Asia, and Africa (http://www.astro4dev.org/blog/category/tf2/visually-impaired/). Since then, the kit has grown to include more tactile materials and activities, with 3D-printable globes of Mercury, Venus, Mars, and the Moon. All the materials are freely available for download and 3D-print from their website, which has become a world beacon for real, practical implementation of the framework of Universal Design for Learning in the space sciences. This framework, adapted to education by Dr. Anne Meyer and Dr. David H. Rose (Meyer et al., 2014) and based on the Universal Design movement in architecture from the 1990s, highlights that modifying design to improve accessibility benefits all users or visitors, not just those with disabilities. Experience with this project has shown that all types of audiences enjoy and learn from participating in the planetarium show with the tactile half-sphere or in the activities with the 3D models (Fig. 2.7).

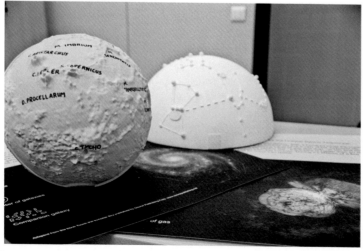

**FIG. 2.7**

3D-printed model of the Moon and celestial half-sphere for planetarium audiences to follow the show in a tactile manner. *Amelia Ortiz-Gil.*

## Concluding Remarks

The space sciences have always been at the forefront of public engagement and outreach of science, not least through science centers and planetariums. The natural beauty of the images and the fascination of the mind-blowing vastness of the universe and everything therein have helped immensely, but communicators have not rested on those laurels and have constantly striven to find new and better ways to reach out to the public. Planetariums and science centers have grown in size and number. Audiences have become larger and more diverse, and the explosion in information and communication technologies has raised the expectations of a public that is more and more exposed and accustomed to high-quality shows and state-of-the-art technology in all areas of life. To keep up the pace of growth and change, new sector-specific technologies have been developed, science communication has been evaluated and researched, and the delivery of outreach now needs to become more and more professionalized.

When I was inspired by Carl Sagan's *Cosmos* (Sagan, 1980) to pursue astrophysics as a career, I could not have imagined that almost 40 years later I would be teaching about such programs and using Sagan as a case study of historic significance in the field, highlighting that, like me, a whole generation of astrophysicists was inspired by his work. Sagan is often referred to as a role model for scientists to get involved in public engagement. He had to struggle with a lack of acceptance by his peers at the time. We can confidently state today that in this regard the public engagement landscape has improved substantially, with funders, universities, and research centers valuing and giving recognition to scientists' involvement in outreach and a community of dedicated science communicators on the rise. Indeed, Carl Sagan was one scientist who decided to devote time to share his passion for astronomy, but *Cosmos* was "a carefully crafted work of rhetorical art" (Schroeder-Sorensen, 2017), professionally designed by a large team of trained specialists in various disciplines.

There has been a period during which the workforce in science centers and planetariums has been made up by enthusiasts and passionate educators and scientists like myself who had little training in the specific field of science communication, let alone in museum studies. I along with many of my colleagues and peers in this line of work were experts in other fields *dedicating* ourselves to science centers and planetariums. In the words of Fernández Navarro (2018), the field is now coming of age and reaching a stage where a transition from a *dedication* to a *profession* is becoming urgent. There are some incipient signs that this transition is taking place. We now have master's degrees in science communication and science museum studies. After gaining extensive practical experience and having had the rare opportunity to complement this experience with academic work in science communication and museum studies, I am glad to be able to contribute to the professionalization through my teaching at a museum studies program that is unique in having a science museum track (http://cgpmuseumstudies.org/). My dream is to see in the not-too-distant future science centers and planetariums become, just as *Cosmos* did in the 1980s, "carefully crafted works of art" created and run by trained specialists in the sector. At the same time, I would not want this to be taken as a call for scientists and researchers to disengage from involvement; on the contrary, this transition can only take place in close collaboration with the scientific community. Communication is all about people, and the accumulated experience of scientists and communicators points in the direction of the human factor: access to scientists and to where science is happening, being able to talk to someone passionate and knowledgeable, and being presented stories about real people and relating the information to one's own lived experience is what draws audiences to truly transformative experiences. For this engagement to continue happening in science centers and planetariums, the support and willingness of scientists and engineers to work hand in hand with a professionalized science communication sector is essential and more necessary than ever. Preparing my science museum studies students for this critical collaboration is part and parcel of my teaching.

## References

Archer, L., Dawson, E., DeWitt, J., Seakins, A., & Wong, B. (2015). "Science capital": A conceptual, methodological, and empirical argument for extending bourdieusian notions of capital beyond the arts. *Journal of Research in Science Teaching, 52*(7), 922–948.

Bourdieu, P. (1986). The forms of capital. In J. Richardson (Ed.), *Handbook of theory and research for the sociology of education* (pp. 241–258). Westport, CT: Greenwood.

Bourdieu, P., & Passeron, J.-C. (1977). *Reproduction in education, society, and culture.* Beverly Hills: CA Sage.

Bruman, R., Hipschman, R., & Exploratorium (San Francisco). (1983). *Exploratorium cookbook: A construction manual for exploratorium exhibits.* San Francisco: The Exploratorium.

Castrillo, A., et al. (2001). *Eclipse Total de Sol, 11 de Agosto de 1999.* Munibe supplement 13. Donostia – San Sebastián, Spain: Sociedad de Ciencias Aranzadi.

Clinard, C. (2019). *UNC's Morehead Planetarium trained 11 of the 12 people who have walked on the moon.* https://abc11.com/science/nasas-first-astronauts-trained-at-uncs-morehead-planetarium/5166118/. (Accessed 8 March 2019).

Falk, J., & Dierking, L. (2010). The 95 percent solution: School is not where most americans learn science. *American Scientist, 98*(6), 486–493.

Fernández Navarro, G. (2018). El Museo de Ciencia Transformador (The transformative science museum). Available at http://www.elmuseodecienciatransformador.org/. (Accessed 19 July 2019).

Hoeg, A. (2012). How to design transformational experiences in out science centers and museums? *The Informal Learning Review, 115*, 10–12.

Hoeg, A. (2017). *Leading Cultural Institutions. 43 recommendations for successful leadership of cultural institutions.* http://media.voog.com/0000/0037/4718/files/LeadingCultural170301.pdf.

Keller, H. (2012). Quo vadis planetarium? *Sterne und Weltraum 8/2012*, 82–87 (in German).

Knappenberger, P., & Duncan, D. (2000). The role of research astronomers in planetaria: Opportunities and dangers. In *Paper given at the conference of the international planetarium society (IPS 2000).*

Leinhardt, G., Crowley, K., & Knutson, K. (2002). *Learning conversations in museums*. Mahwah, N.J.: Lawrence Erlbaum.

Louden, J., Wilson, P., & ad Nollenberg, J. (2016). Searching the outer solar system for transneptunian objects. In *Poster presented at the 2016 student research and creativity day*. NY: SUNY Oneonta. Available at http://www.oneonta.edu/academics/research/PDFs/SRCA2016-Louden.pdf. (Accessed 19 July 2019).

Meyer, A., Rose, D. H., & Gordon, D. (2014). *Universal design for learning: Theory and practice*. Wakefield, MA: CAST.

Minguez, J., et al. (2004). *Luces del Norte*. Munibe supplement 18. Donostia — San Sebastián, Spain: Sociedad de Ciencias Aranzadi.

Pine, B. J., & Gilmore, J. H. (1999). *The experience economy*. Boston, Mass: Harvard Business Review Press.

Ratcliffe, M. (1996). Should planetarium shows Be live or prerecorded? *Mercury*, 10. March-April 1996.

Sagan, C. (1980). *Cosmos (TV series). Produced by Gregory Andorfer*.

Schirrmacher, A. (2018). The foundations of interactive exhibit design: An historical discussion. In *Presentation at the 2018 annual conference of the association of science -technology centers (ASTC2018)*.

Schroeder-Sorensen, K. (2017). *Cosmos and the rhetoric of popular science*. Lanham, MD: Lexington Books.

Science Museum Group. (2018). *Engaging all audiences with science. Science capital and informal science learning*. Available at https://group.sciencemuseum.org.uk/wp-content/uploads/2018/10/science-museum-group-engaging-all-audiences-with-science.pdf. (Accessed 21 July 2019).

Wagensberg, J. (2006). Towards a total museology through conversation between audience, museologists, architects and builders. In M. Arnal (Ed.), *The total museum* (pp. 11−103). Barcelona, Spain: Sacyr SAU.

# Engaging the Public With the Great American Eclipse of 2017

3

C. Alex Young, Shannon P. Reed

NASA has excited and engaged the public since its beginnings at the start of the space age in 1958. During the space race of the 1960s with the Mercury, Gemini, and Apollo programs, NASA held the public captive with fascination about space. The peak of this captivation was arguably during Apollo 11 and the first human setting foot on the Moon. Other times with similar interest for human space flight occurred with the start of the Space Shuttle program in 1980.

More excitement was provided with robotics missions as NASA beamed back the first space images of the outer planets with the Pioneer and Voyager spacecraft and then the most distance objects in the universe with the Hubble Space Telescope. These moments of sharing exploration and science are part of the very fundamental tenets of NASA, whose organic legislation directs the agency both to pursue the "expansion of human knowledge of the Earth and of phenomena in the atmosphere and space …" and to "provide for the widest practicable and appropriate dissemination of information concerning its activities and the results thereof" (National Aeronautics and Space Administration, 1958).

Thus, woven into NASA's fabric is the promise to share all that it does with the public. This was simply the mandate of a civilian government agency to provide transparency to the public. The agency soon realized that its exploration efforts and its science objectives naturally excited and engaged the public. The curiosity of the public and its fascination with the unknown was fueled by the fundamental work of NASA. Also, NASA had a goal to learn about Earth and space. The organization and the people of NASA were adding to our understanding; they were educating the world through their work. And the public clearly wanted more, as did educators and students. NASA is not a formal education organization, but it was adding to the current state of knowledge. Educators and students were interested in sharing and connecting with this exciting new information. Members of NASA, from scientists and engineers to management, also felt it part of their responsibility to support public outreach. Outreach thus formally became part of NASA's core goals, part of its strategic objectives to "share the story, the science, and the adventure of NASA missions and research to engage the public in scientific exploration and contribute to improving science, technology, engineering, and mathematics (STEM) education nationwide" (National Aeronautics and Space Administration, 2018).

Space Science and Public Engagement. https://doi.org/10.1016/B978-0-12-817390-9.00001-4

The internal structure and organization of NASA continues to evolve and change to address its goals, but the fundamental objective has been to share its science and exploration through public engagement as well as to provide resources to promote and support STEM education across the country. Through its many direct public engagements in its quest to explore our universe with humans, robotic spacecraft, aircraft, balloons, and rockets, NASA has given us moments like the "seven minutes of terror" with the landing of the Curiosity and Perseverance rovers on Mars, a flyby of Pluto with New Horizons, and a look at another star system (actually thousands of them) with the Kepler spacecraft. All of these and thousands of other moments have provided the nation and the world an opportunity to glimpse at nature's marvels through the eyes of NASA. NASA has been mostly sharing nature through its eyes, i.e., its telescopes and detectors in space. The beginnings of astronomy and space science come from what we see here from Earth and not just what we can see from space. While NASA has opened up a new and exciting way to view the universe, it is important to remember the visceral connection with space we all get when we look at something above with our own eyes (either with the naked eye or telescopes). Humanity has always enjoyed looking at the stars or the Moon and the rarer event of a total solar eclipse (see Fig. 3.1).

Though not completely appreciated at the time by all, a unique opportunity to engage the public came to NASA in the form of a total solar eclipse on August 21, 2017. A solar eclipse happens when the Moon casts a shadow on Earth, fully or partially blocking the Sun's light in some areas. The result is a total or partial eclipse, respectively (see Fig. 3.2).

FIG. 3.1

Composite image of 2017 total solar eclipse from Wyoming, United States. *Antonis Farmakopoulos.*

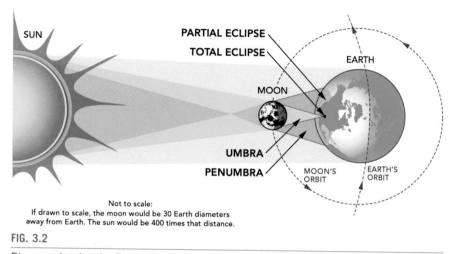

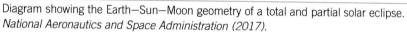

**FIG. 3.2**

Diagram showing the Earth–Sun–Moon geometry of a total and partial solar eclipse. *National Aeronautics and Space Administration (2017).*

NASA scientists have always been interested in eclipses, and NASA has provided important information to the public and the science community for many years through the eclipse.gsfc.nasa.gov website created by Dr. Fred Espenak, a retired NASA astrophysicist. The 2017 eclipse, however, presented a special opportunity for the national agency, as it was the first total solar eclipse visible in the contiguous United States in 38 years. This was also the first total solar eclipse crossing only the United States since 1918. The path of totality (the path of the umbral shadow of the Moon, within which the Sun's visible disk is completely blocked out) crossed 14 states (see Fig. 3.3).

The states in the path of totality were Oregon, Montana, Idaho, Wyoming, Nebraska, Iowa, Kansas, Missouri, Illinois, Kentucky, Tennessee, Georgia, North Carolina, and South Carolina. Outside the path of totality, the rest of the continental United States was within the Moon's penumbral shadow, where the Moon only partially blocked the Sun and created a partial solar eclipse.

## A Plan to Support the Eclipse

NASA made a decision in September 2016 that the agency would support the nation's experience with the 2017 total solar eclipse. The agency's Science Mission Directorate (SMD), which supports research on the Sun, Earth, Moon, solar system, and universe, began coordinating activities across NASA's 10 field centers and with its partners in October 2016 to avoid duplication and ensure awareness of each other's activities. Many individuals and organizations had been thinking about and planning for this eclipse for years, some even decades. NASA's goal was to

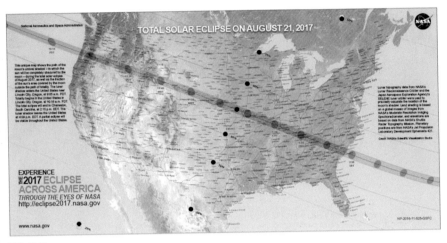

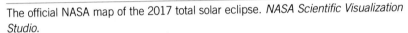

**FIG. 3.3**

The official NASA map of the 2017 total solar eclipse. *NASA Scientific Visualization Studio.*

use its resources and leverage its brand to support and amplify all of the great work and plans already in place or being developed. In particular, NASA wanted to share its unique content and science. NASA organized all of its public information within the NASA 2017 eclipse website (National Aeronautics and Space Administration, 2017).

## Safety

Under the leadership of SMD and with the support of one of its science education teams (the NASA Space Science Education Consortium (NSSEC, formerly called the Heliophysics Education Consortium (Young, Debebe, et al., 2019, p. 287))), the NASA team developed a plan for the eclipse based around five main focus areas. As the principal investigator (Young) and program manager (Reed) of the NSSEC, we led the primary NASA team. One of us (Young) is also a solar physicist and served as one of the primary science experts and eclipse spokespersons for NASA. The first and most important area that NASA focused on was safety. Most of us have heard the message since childhood to never look—and more importantly never stare—at the Sun because it could damage your eyes. This warning also applies to solar eclipses. The light from the Sun is so intense that looking at it for more than a brief glance runs the risk of damage (possibly permanent damage) to your retina, causing partial or total loss of eyesight. Only during the brief period of totality when only the Sun's corona, or outer atmosphere, is visible can the eclipse be viewed with the unprotected eye. During this short period, viewable within the eclipse's path of totality, the eclipse is about as bright as a full moon. Otherwise, during the partial phase of the eclipse, both within and outside the path of totality, you can never look directly at the Sun without safe solar viewing glasses.

NASA and partners distributed 3.3 million pairs of ISO 12312-2 certified safe solar viewing glasses. ISO 12312-2 is the international standard for safe solar viewing glasses established in 2015 (International Organization for Standardization, 2015). Of the 3.3 million pairs of glasses distributed by NASA and its partners, 500,000 were printed by NASA. Another 2.1 million were printed by Google and distributed by the Moore Foundation, Google, the Research Corporation, and NASA to 7100 public libraries that were unable to purchase their own glasses (Dusenbery et al., 2018; Fraknoi et al., 2019, p. 241). Safe viewing techniques were highlighted including safe methods for direct telescopic observations, pinhole projectors, and "Sun funnel" assembly and use instructions (National Aeronautics and Space Administration, 2017; Zych, 2019, p. 229).

Concern for safety was not limited to safe solar viewing but also included event and travel safety (Parvinashtiani et al., 2019). NASA and its partners encouraged people to travel to the path of totality if at all possible. Approximately 12 million people live within the path of totality, and it was also estimated that two-thirds of the US population live within about a day's drive (500 miles or 800 km) (Zeiler, 2017). A surge in traffic was expected for the eclipse. There were also hundreds of events planned both within and outside the path. NASA gathered or pointed to previously gathered guidance for travel safety and event safety. The eclipse occurred on a Monday (standard workday) and was the first day of school for many K-12 schools as well as a move-in day for many colleges and universities. The roads were expected to be busy already on top of the possible millions traveling short and long distances to view the eclipse. As a day in mid-August, it was anticipated to be hot and humid in many locations along with path. NASA provided information on safety and preparedness for eclipse events through its websites, social media, conferences and workshops, and printed materials. Additionally, every official NASA eclipse event was required to have a safety plan in place.

NASA worked with the American Astronomical Society (AAS) to craft a joint solar viewing safety message for the public (Fienberg, 2019, p.391). This message was vetted by the National Science Foundation, the American Academy of Optometry, the American Optometric Association, the Canadian Association of Optometrists, and the American Academy of Ophthalmology. This message formed a safety bulletin (see Fig. 3.4), which was posted on both AAS and NASA Eclipse 2017 websites and was translated to Spanish. The NASA 2017 eclipse website safety page drew 10.9 million-page views in August 2017.

The marketplace for eclipse glasses was flooded with counterfeit glasses in the weeks leading up to eclipse. A large portion of these came from China and appeared on Amazon.com and other online retailers. In response to concerns about the safety of individual solar viewing glasses and the fear of unsafe glasses, NASA directed users to the AAS website detailing vetted producers and distributers of ISO-compliant solar viewing glasses (Fienberg, n.d.).

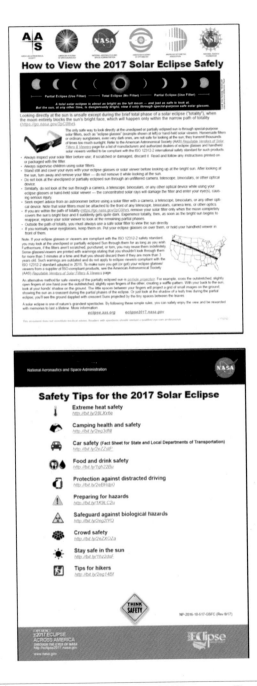

**FIG. 3.4**

Snapshots of the eclipse safety pages from eclipse2017.nasa.gov. The first page covers viewing safety, and the second covers event safety. *National Aeronautics and Space Administration (2017).*

## Science and Citizen Science

NASA is a science organization. Its goal to support the 2017 eclipse also included enabling and supporting science happening during the eclipse. Part of that science focused on studying the Sun's outer atmosphere, the corona. Even when there is not an eclipse, the corona can be studied by a special camera called a coronagraph. It creates an artificial eclipse with a device in the telescope that blocks out the bright solar disk. NASA flies these coronagraphs in space to give close to a 24/7 view of the corona. The downside is that coronagraphs do not reveal the corona as close to the surface of the Sun as total eclipses created by the Moon. Because total solar eclipses provide a unique view of the corona, NASA funded observations of the corona during the total eclipse from the ground and airplanes. NASA also funded experiments to study Earth, including land and atmospheric responses. Total solar eclipses provide an opportunity to study Earth under uncommon conditions. The sudden blocking of the Sun during an eclipse reduces the light and temperature on the ground, and these quick-changing conditions can affect weather (Papol, 2019), vegetation, and animal behavior (Heywood et al., 2019, p. 473; Sudbrink et al., 2019, p. 457; Young, Johnson et al., 2019, p. 463). A key piece of NASA's support was the unique set of space-based and high-altitude observations that NASA provided with its resources. A collection of 11 spacecraft from NASA and partner organizations—the National Oceanic and Atmospheric Administration (NOAA), the Japanese Aerospace Exploration Agency (JAXA), and the European Space Agency (ESA)—provided observations of the Sun, Moon, and Earth during the eclipse. In addition, NASA provided a Gulfstream-III and two WB-57 aircraft for high-altitude observations. The International Space Station provided a unique view with photographs of the eclipse shadow moving across the United States (see Fig. 3.5).

The Heliophysics Division of SMD funded 11 teams to study the Sun and Earth during the eclipse, and some of the research included testing new equipment for future ground- and space-based observations during eclipses. Some of the data collected during the eclipse are still being analyzed, but the preliminary results have been published in numerous scientific papers and partially summarized (Young, 2018).

The science that NASA supported was not purely conducted by professionals but also included student- and amateur-supported citizen science. Citizen science is a powerful way to further science, engage the public, and promote scientific literacy (see Fortson's chapter on citizen science for further discussion). The eclipse provided a wonderful opportunity to support citizen science across the country and support many different areas of research. NASA supported the Citizen CATE telescope program (Penn et al., 2020), which trained operators to maintain 65 observing sites along the path of totality, providing a more than 1-hour eclipse totality observation by combining these observations along the path. A high-altitude balloon program led by the Montana Space Grant Consortium (Des Jardins et al., 2019, p. 353; McCracken et al., 2019, p. 491; Sibbernsen & Sibbernsen, 2019, p. 485) had more than 50 teams across the United States providing a unique view of the shadow and a rich set of observations of atmospheric changes due to the eclipse's shadow.

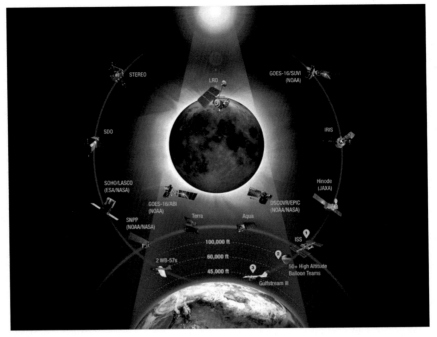

**FIG. 3.5**

An illustration of the air and space assets (11 spacecraft, aircraft, high-altitude balloons, and the International Space Station) providing observations during the 2017 total solar eclipse. The assets marked with a 1, 2, or 3 were also part of the broadcast locations supporting web and TV broadcasts during the event. *NASA.*

NASA also supported several programs that used mobile apps to make eclipse observations with the help of citizen scientists. The most notable of these is GLOBE Observer (Rahman et al., 2019, p. 501; Weaver et al., 2019, p. 511), which provided temperature and cloud observations by users from coast to coast.

## Education and Public Engagement

While safety was the most important focus and science is what NASA is all about, the NASA-led eclipse program centered on education and public engagement around the eclipse. This event was a unique opportunity to help the public share in the excitement and wonder while developing resources for the public, many of which also supported education. This was an amazing teachable moment supported by many government agencies in addition to NASA, such as the US Fish and Wildlife Service, National Park Service, National Oceanic and Atmospheric Adminstration, Department of Transportation and Department of Education. Outside of governments, organizations involved included zoos, schools, libraries, cities and towns, baseball parks, and many more.

Science pervades a huge portion of public life, not just the lives of the scientists and technologists. Traditionally, scientists and science organizations have tried to educate (tell) the public about their work with a one-sided, authoritative approach, but this approach has not been fully successful. A newer approach, public engagement, developed to involve the public in a more open, genuine, and multidirectional dialogue. This serves in the development of mutual respect but also an honest discussion of science—the good and the bad, the successes and the failures. This not only leads to better understanding by the public of science but also the scientists' understanding of the where science fits and connects with society. This open conversation also helps the public understand the process of science, which is far richer and more complex than what is often taught to us in school.

Here, we are using the term "education" to mean a more formal connection with educators and students. Education is similar to public engagement, but there are also specific learning goals and desired learning outcomes. Education is roughly put into two areas: formal (in the classroom) and informal (outside the classroom). Some very specific locations for informal learning include libraries and museums.

As the premier space science agency in the world, NASA is involved in leading-edge research on the Sun, Earth, Moon, solar system, and beyond. NASA brought this enormous capability to bear on the 2017 eclipse to engage, excite, and educate the country through its many scientists and education specialists, interviews, publications, online materials, social and broadcast media, conference talks and workshops, and print materials. Through these venues, NASA sought to improve the understanding of eclipses, how they work, and, ultimately, the overall science literacy of the country (Miller, 2017), especially for traditionally underserved populations. NASA emphasized the connections between the many divisions of space and Earth science specialties and the interconnectedness of our many programs of research, exploration, and discovery. Though the science that NASA supported focused on the Sun, Moon, and Earth, it was an opportunity to highlight the connection between eclipses on Earth and transits of exoplanets (Lawton et al., 2019, p. 307).

NASA provided science education articles, presentations, and materials that told the story of eclipse science to the country. NASA distributed 20,000 eclipse packets (folders with glasses, posters, and several information flyers) to educators and tens of thousands of people. NASA included a section on the NASA eclipse website entitled "Eclipse 101" that detailed the history of eclipses and their impact on society and the basic geometry and celestial mechanics that produces eclipses and allows us to predict them not only here on Earth but throughout the solar system and beyond. The website also included "Eclipse Who What When Where How," a treatise on the eclipse describing where one should be, what one would see, how to view it, misconceptions, and more. NASA scientists gave presentations and participated in workshops and panel discussions at numerous science and education conferences and public events and responded to questions from the general public through the website. NASA members wrote articles for major science education publications such as *Sky and Telescope* and sat for interviews for media outlets such as WGBH, Science Friday, and Planetary Radio. NASA subject matter experts (SMEs) consulted on the science content for eclipse activities and print materials.

NASA solicited the help of scientists, amateur astronomers, and science educators across the country to communicate the science of the eclipse to millions. Officially, over 300 SMEs registered to communicate NASA's messages, but unofficially thousands helped out in various eclipse events. Each was vetted for sufficient content knowledge and availability. Training materials on eclipse science, NASA missions, viewing techniques, and safety precautions were provided to each SME.

To engage the masses both within and outside of the path of totality, NASA supported and promoted events across the country and provided maps to enable the public to find events near them. These included NASA official events, defined as having met certain requirements for safety and SME support, as well as NASA general events. These even included events leading up to the day of the eclipse, and all were recorded using Google maps (Fig. 3.6). All of this information along with the supporting information about SMEs, safety, and available materials was provided on the NASA eclipse website.

Libraries and museums (Libraries; McCarthy & Jackson, 2019, p. 313) played an especially important role helping to engage the public by providing information, connections to SMEs, viewing events, and glasses. NASA helped to support some of this through the NASA @ My Library program and the 7100 STAR Net libraries network. Both of these initiatives aim to promote science learning opportunities through public libraries.

Through social media, NASA reached more than 3.6 billion (nonunique) users, and Twitter reports that there were more than 6 million eclipse Tweets on eclipse day. Social media proved a powerful tool to engage and connect the public with

FIG. 3.6

These two maps show the NASA official event and general event locations as identified on the eclipse2017.nasa.gov website. *National Aeronautics and Space Administration (2017).*

eclipse content from NASA. The NASA social media team used some creative posts to connect with specific audiences. For example, the @NASAMoon Twitter account "blocked" the @NASASun account on eclipse day. It was the most shared Tweet and gained over 27,000 new followers for the @NASAMoon account. The agency's most popular Instagram image ever was posted on eclipse day and received over 1.4 million likes. NASA accounts gained a significant number of followers on Twitter, Facebook, and Instagram on eclipse day. This meant that NASA content and posts continued to show up in followers' news feeds well beyond the eclipse. The NASA HQ Photo Flickr account saw more than 1.5 million photo views on August 21, 20–30 times higher than daily averages earlier in the month.

Based on past events NASA covered, websites were expected to see a large amount of traffic. NASA's experience covering the 2012 transit of Venus, for example, was an important indicator to determine how much website traffic to expect on the day of the eclipse as well as during the time leading up to the event. NASA expected a large amount of web traffic, but the amount received far exceeded those expectations. The nasa.gov/eclipselive site and the official NASA 2017 eclipse site received more than 90 million-page views on August 21, topping NASA's previous web traffic record about seven times over. The NASA 2017 eclipse site was the top government agency site for the 2 weeks leading up to and during the eclipse, according to analytics.usa.gov.

In addition to page views, the NASA 2017 eclipse site recorded many downloads of its products. The NASA map of the United States that showed the path of totality (Fig. 3.2) was the single most popular printed item and was downloaded more than 4 million times.

NASA teams and partners worked hard to address the needs and requirements of underserved audiences whenever possible. Such audiences included but were not limited to the blind and visually impaired, indigenous communities, and Spanish-speaking communities. NASA promoted and shared a Braille book, *Getting a Feel for Eclipses*, and supported the development of the Eclipse Soundscapes smartphone/tablet application (Eclipse soundscapes application, n.d). The tactile book allowed the visually impaired to feel the orbital mechanics and progression of a solar eclipse and the path of totality for the 2017 eclipse across the United States. The Eclipse Soundscapes application provided a multisensory experience allowing users to hear and feel eclipse imagery by creating unique "rumble maps" that translated color and intensity of images into sound and vibrations (see Fig. 3.7).

The Indigenous Education Institute (IEI) produced videos on Indigenous eclipse teachings with the Cherokee Nation and the Navajo Nation. Through a partnership with the University of Northern Texas, IEI used not only the safe-viewing solar glasses but also 3D-printable eclipse files for Navajo and Cherokee to engage Indigenous communities. The University of Alaska in Fairbanks held multiple eclipse events to engage the Alaska Natives, especially in Ketchikan, Alaska.

NASA Television conducted a 4-hour live broadcast, "Eclipse Across America: Through the Eyes of NASA," to provide coverage before, during, and after the

**FIG. 3.7**

This collage highlights the development of Eclipse Soundscapes along with its use by the blind and visually impaired public. *Top left panel, Kelsey Perrett; remaining panels, Carroll Center for the Blind.*

eclipse. A preview show was hosted from Charleston, South Carolina. The main show began at 1 p.m. EDT and covered the path of totality the eclipse took across the United States, from Oregon to South Carolina (see Fig. 3.8). The program featured views from NASA research aircraft, high-altitude balloons, satellites, and specially modified telescopes. It also included live reports from Charleston as well as from Salem, Oregon; Idaho Falls, Idaho; Beatrice, Nebraska; Jefferson City, Missouri; Carbondale, Illinois; Hopkinsville, Kentucky; and Clarksville, Tennessee. In addition to showing video of the eclipse, the broadcast took viewers to activities taking place in national parks, libraries, stadiums, and museums across the nation (Baer, 2019, p. 23; Davis et al., 2019, p. 365; Wasser et al., 2019, p. 321).

## What is next?

August 21, 2017, was an amazing day! Thanks to the efforts of many groups, millions of people had an opportunity to safely experience an eclipse in either the partial or total phase. It is estimated that around 215 million Americans viewed that eclipse:

**FIG. 3.8**

NASA broadcast locations which supported the 4-hour live broadcast. Then was hosted at location 14. Locations 1, 2, and 3 are shown in Fig. 3.5. *NASA*.

61 million saw it electronically and 154 million stepped outside to see it with their own eyes (Miller, 2017). The number who saw totality is at least 12 million and it could be more. In addition, millions from the rest of the world watched the eclipse via the live broadcasts. All of these views make the 2017 Great American Eclipse arguably the most observed eclipse in history, and it is estimated to be one of the most viewed events in NASA's history short of only the Apollo 11 Moon landing.

There is more excitement on the horizon because we have more eclipses in our future. Total solar eclipses happen on average every 18 months somewhere on Earth, and this will continue for millions of more years. But Earth is very big, meaning that on average they cross any point only once every 450 years. This is what makes them rare. There was a total solar eclipse in Chile and Argentina in 2019, and there will be another one farther south in Chile and Argentina in 2020. The next total solar eclipse after that is even farther south, in Antarctica in 2021. But then the United States will see the next solar eclipses, with an annular eclipse in 2023 and another total solar eclipse on April 8, 2024. The 2024 eclipse starts on land in Mexico, reaches the United States in Texas, leaves the United States in New England, and leaves North America in Canada (see Fig. 3.9). We already had a total solar eclipse across the United States in 2017, and we will have another one only 7 years later. The two

**FIG. 3.9**

A map of the total solar eclipse of April 8, 2024. *NASA Scientific Visualization Studio.*

eclipses' path even cross, around Carbondale, IL. This makes them both even more exciting and unique for the continental United States.

The 2024 eclipse could prove to be even bigger than the 2017 eclipse: the path of totality is nearly twice as wide, the average duration of totality is nearly twice as long, and the path covers many more populated areas and cities. All of NASA's and others' public engagement has paid off. We now have a country with a lot more people who want to see another total solar eclipse or see what they missed in 2017.

Many groups from across the country and around the world participated in the eclipse, supporting science and engaging the public. NASA was a large part of this, but it took work from many other public, private, and government groups. All played an important role in some fashion, and many of these groups provided lessons learned about what went well as well as what might be done differently or better in the future.

We learned that there was a desperate need for glasses as well as up-to-date information on ensuring the glasses were from a reputable dealer. The partial eclipse lasted for hours, and it was most likely that no one was going to look at the partial eclipse for the full time period. So it is important to better convey that glasses can be shared. The eclipse community will need to do a better job explaining that being in the path of totality when totality occurs, compared to just off the path even during the time of totality is quite literally like night and day. There is no partial totality, and there is no "higher degree of totality" (Fienberg, 2019, p. 391). While it is important to support and engage even those who could not reach totality, we do not want people to travel to just outside of the path of totality because they did not realize the difference and miss out on the experience of totality. One of the most powerful take-homes was the thirst for more even after August 21, 2017. Social media showed us that even after we engaged the public, enjoyed the eclipse, and went home, the interest online continued for days after the event. The engagement community was not prepared to meet that need. These are important lessons to take with us to 2024 and beyond.

The 2017 eclipse brought together people from all walks of life across the country in a single focus. They all had the chance to connect with space science in a personal way through the power of a total solar eclipse. Nature came to us and allowed us to get a glimpse of our amazing universe. NASA and many others will be engaging the world again with what we might end up calling the "Great North American Eclipse." Get your glasses and be amazed.

# References

Baer, R. (June 2019). Eclipse 2017 at Southern Illinois University Carbondale. In S. R. Buxner, L. Shore, & J. B. Jensen (Eds.), *Celebrating the 2017 Great American Eclipse: Lessons Learned from the Path of Totality, ASP Conference Series* (Vol. 516, p. 23). San Francisco, CA.

Davis, A., Cook, S., Engler, M., Jones, A., Melena, S., Olson, J. G., & Paglierani, R. (June 2019). See you in the shadow: A national parks and NASA collaboration. In S. R. Buxner, L. Shore, & J. B. Jensen (Eds.), *Celebrating the 2017 Great American Eclipse: Lessons Learned from the Path of Totality, ASP Conference Series* (Vol. 516, p. 365). San Francisco, CA.

Des Jardins, A. C., Mayer-Gawlik, S., Larimer, R., Knighton, W. B., Fowler, J., Ross, D., Koehler, C., Guzik, T. G., Granger, D., Flaten, J., & Grimber, B. I. (June 2019). Eclipse ballooning project live streaming activity: Overview, outcomes, and lessons learned. In S. R. Buxner, L. Shore, & J. B. Jensen (Eds.), *Celebrating the 2017 Great American Eclipse: Lessons Learned from the Path of Totality, ASP Conference Series* (Vol. 516, p. 353). San Francisco, CA.

Dusenbery, P., et al. (2018). *Eclipse 2017: A celestial achievement for public libraries.* Space Science Institute. https://www.starnetlibraries.org/2017eclipse/.

*Eclipse soundscapes application.* Retrieved from https://eclipsesoundscapes.org.

Fienberg, R. T., *AAS offers updated advice for safely viewing the solar eclipse.* Retrieved from https://aas.org/media/press-releases/aas-offers-updated-advice-safely-viewing-solar-eclipse.

Fienberg, R. T. (June 2019). What the American Astronomical Society Learned from the "Great American Eclipse". In S. R. Buxner, L. Shore, & J. B. Jensen (Eds.), *Celebrating the 2017 Great American Eclipse: Lessons Learned from the Path of Totality, ASP Conference Series* (Vol. 516, p. 391). San Francisco, CA.

Fraknoi, A., Schatz, D., Duncan, D., Dusenbery, P., Holland, A., LaConte, K., & Mosshammer, G. (June 2019). Eclipse 2017 in libraries: The impact of providing 2.1 million eclipse glasses and support to 7,100 libraries nationwide. In S. R. Buxner, L. Shore, & J. B. Jensen (Eds.), *Celebrating the 2017 Great American Eclipse: Lessons Learned from the Path of Totality, ASP Conference Series* (Vol. 516, p. 241). San Francisco, CA.

Heywood, N. C., Zellmer, P., & Reser, R. (June 2019). Birds, bees—and cookies? Great plains bioclimatology observations during the total solar eclipse of 21 August 2017. In S. R. Buxner, L. Shore, & J. B. Jensen (Eds.), *Celebrating the 2017 Great American Eclipse: Lessons Learned from the Path of Totality, ASP Conference Series* (Vol. 516, p. 473). San Francisco, CA.

International Organization for Standardization. Technical committee 94, subcommittee 6 2015, ISO 12312-2:2015: Eye and face protection − sunglasses and related eyewear − Part 2: Filters for the direction observation of the Sun. Retrieved from https://www.iso.org/standard/59289.html, 2015.

Lawton, B., Smith, D., Squires, G., Biferno, A., Lestition, K., Cominsky, L., Rhue, T., Slivinski, C., Godfrey, J., Arcand, K., Dussault, M., & Winter, H. D. (June 2019). Connecting the Great American Eclipse to NASA astrophysics. In S. R. Buxner, L. Shore, & J. B. Jensen (Eds.), *Celebrating the 2017 Great American Eclipse: Lessons Learned from the Path of Totality, ASP Conference Series* (Vol. 516, p. 307). San Francisco, CA.

McCarthy, C., & Jackson, A. (June 2019). Engaging museum and science center audiences across the United States during the 2017 solar eclipse. In S. R. Buxner, L. Shore, & J. B. Jensen (Eds.), *Celebrating the 2017 Great American Eclipse: Lessons Learned from the Path of Totality, ASP Conference Series* (Vol. 516, p. 313). San Francisco, CA.

McCracken, M., Critchlow, D., Hill, Z., Sencabaugh, M., Robertson, J., Gaither, B., & Oelgoetz, J. (June 2019). High altitude ballooning using amateur equipment: A primer. In S. R. Buxner, L. Shore, & J. B. Jensen (Eds.), *Celebrating the 2017 Great American Eclipse: Lessons Learned from the Path of Totality, ASP Conference Series* (Vol. 516, p. 491). San Francisco, CA.

Miller, J. D. (2017). *Americans and the 2017 eclipse: An initial report on public viewing if the August total solar eclipse.* University of Michigan. http://ns.umich.edu/new/releases/25108-a-record-number-of-americans-viewed-the-2017-solar-eclipse.

National Aeronautics and Space Administration. (2017). *Eclipse website.* NASA. https://eclipse2017.nasa.gov.

*NASA @ My Libraries Program.* Retrieved from https://www.starnetlibraries.org/.

National Aeronautics and Space Administration. (1958). *Space Act.* NASA.

National Aeronautics and Space Administration. (2018). *Strategic Plan.* NASA.

Papol, A. (June 2019). Effects of the Great American Eclipse on Weather Phenomena Across the Contiguous United States. In S. R. Buxner, L. Shore, & J. B. Jensen (Eds.), *Celebrating the 2017 Great American Eclipse: Lessons Learned from The Path of Totality, ASP Conference Series* (Vol. 516, p. 421). San Francisco, CA.

Parvinashtiani, N., Hopps, A., & Radow, L. (June 2019). Transportation planning and preparation for 2017 solar eclipse. In S. R. Buxner, L. Shore, & J. B. Jensen (Eds.), *Celebrating the 2017 Great American Eclipse: Lessons Learned from The Path of Totality, ASP Conference Series* (Vol. 516, p. 269). San Francisco, CA.

Penn, M. J., et al. (2020). Acceleration of coronal mass ejection plasma in the low corona as measured by the citizen CATE experiment. *Publications of the Astronomical Society of the Pacific, 132*(1007), 014201.

Rahman, M. I., Czajkowski, K., Jiang, Y., & Weaver, K. (June 2019). Validation of GLOBE citizen science air temperature observations using data from the great American solar eclipse. In S. R. Buxner, L. Shore, & J. B. Jensen (Eds.), *Celebrating the 2017 Great American Eclipse: Lessons Learned from The Path of Totality, ASP Conference Series* (Vol. 516, p. 501). San Francisco, CA.

Sibbernsen, K., & Sibbernsen, M. (June 2019). Viewing the eclipse from 109,000 feet and the road to get there. In S. R. Buxner, L. Shore, & J. B. Jensen (Eds.), *Celebrating the 2017 Great American Eclipse: Lessons Learned from The Path of Totality, ASP Conference Series* (Vol. 516, p. 485). San Francisco, CA.

Sudbrink, D. L., Mills, R., Moore, R., & Rendleman, E. (June 2019). Incorporating students into investigations of the effects of solar eclipse totality on biological organisms. In S. R. Buxner, L. Shore, & J. B. Jensen (Eds.), *Celebrating the 2017 Great American Eclipse: Lessons Learned from The Path of Totality, ASP Conference Series* (Vol. 516, p. 457). San Francisco, CA.

Wasser, M., Jones, A., Petro, N., & Bleacher, L. (June 2019). Total eclipse of the ballpark: NASA's lunar reconnaissance orbiter team's partnership with Minor league Baseball. In S. R. Buxner, L. Shore, & J. B. Jensen (Eds.), *Celebrating the 2017 Great American Eclipse: Lessons Learned from The Path of Totality, ASP Conference Series* (Vol. 516, p. 321). San Francisco, CA.

Weaver, K., Kohl, H., Martin, A., & Burdick, A. (June 2019). How cool was the eclipse? Collecting Earth science data with citizen scientists and GLOBE observer. In S. R. Buxner, L. Shore, & J. B. Jensen (Eds.), *Celebrating the 2017 Great American Eclipse: Lessons Learned from The Path of Totality, ASP Conference Series* (Vol. 516, p. 511). San Francisco, CA.

Young, C. A. (August 2018). *Shadow science*. Sky & Telescope.

Young, C. A., Debebe, A., Cline, T., Lewis, E., Mayo, L., Ng, C., Odenwald, S., Reed, S., Sasser, L., & Stephenson, B. (June 2019). Eclipse 2017: Through the eyes of NASA. In S. R. Buxner, L. Shore, & J. B. Jensen (Eds.), *Celebrating the 2017 Great American Eclipse: Lessons Learned from The Path of Totality, ASP Conference Series* (Vol. 516, p. 287). San Francisco, CA.

Young, A. N., Johnson, R. F., Ricard, E., & Wyatt, R. (June 2019). Life responds: Citizen science to document behavior changes in plants and animals during the 2017 total solar eclipse. In S. R. Buxner, L. Shore, & J. B. Jensen (Eds.), *Celebrating the 2017 Great American Eclipse: Lessons Learned from The Path of Totality, ASP Conference Series* (Vol. 516, p. 463). San Francisco, CA.

Zeiler, M. (2017). *Predicting eclipse visitation with population statistics*. Great American Eclipse. https://www.greatamericaneclipse.com/statistics/.

Zych, A. (June 2019). A just-in-time strategy for sharing eclipse viewing information the surprise success of facebook live for sharing safe viewing alternatives to solar eclipse glasses. In S. R. Buxner, L. Shore, & J. B. Jensen (Eds.), *Celebrating the 2017 Great American Eclipse: Lessons Learned from The Path of Totality, ASP Conference Series* (Vol. 516, p. 229). San Francisco, CA.

# The Power of Hubble Space Telescope Imagery

**Zolt Levay**

## Introduction

Astronomy has always been a visual activity. People were trying to understand the wonders of the night sky well before any instruments enhanced our naked-eye view. Galileo Galilei changed all that with his use of the newly invented telescope to enhance the view of the mysterious lights in the sky by amplifying the feeble light, instantly expanding our understanding of the universe. The advent of photography allowed us to faithfully record the view from increasingly powerful telescopes and to acquire more precise, quantitative measurements. But photography also dramatically extended our view of the deep universe by being able to record light much fainter than our eyes are able to see and enabled the reproduction of images for wider distribution, first through printing. The development of digital imaging provided another leap in the ability to probe the deep universe and also facilitated the nearly instantaneous, worldwide distribution of images and other content to inform an unprecedentedly diverse audience about our ever-expanding understanding of the universe and our place in it.

Images have captured human imagination, from early adornments and rock art to painting and photography. Astronomy in general captures the imagination of many people partly because we all can relate to the night sky at some level, whether it is by seeing the Moon occasionally or gazing in awe of the Milky Way arching across a dark, star-filled sky. Powerful telescopes produce dramatic images of the sky that may appear abstract to the uninitiated but in fact are entirely representational photographs of real, dynamic landscapes, albeit incomprehensibly far away.

The Hubble Space Telescope in particular has been producing dramatic images of the deep cosmic landscape for several decades. It represents a "case study" of the confluence of circumstances to propel an enterprise into the public's awareness—in this case an astronomical observatory. Hubble is one of a constellation of astronomical facilities working together to further our understanding of the universe. Many world-class observatories are supporting cutting-edge science and producing dramatic images, and even backyard astrophotographers are producing amazing images and contributing to research. Nevertheless, Hubble remains foremost in the identification of astronomy by most people. Many factors have led to Hubble being so widely recognized among the public. The dramatic history of the Hubble—founded from a bold vision with grand expectations, declared a failure after launch, redeemed

**Space Science and Public Engagement. https://doi.org/10.1016/B978-0-12-817390-9.00010-5**

with risk-defying human space flight, and ultimately producing a steady stream of revolutionary science and compelling images—cemented the mission in the public consciousness as *the* most important astronomy facility today, if not for all time.

As someone who was associated for more than 35 years with Hubble and the Space Telescope Science Institute (STScI) in Baltimore, Maryland, that operates it, spending most of that time producing images from Hubble data, I examine how Hubble opened the universe to the global public in ways never before possible. The lessons learned about the reach of powerful imagery through the experiences of Hubble's long mission should be applicable to the development of operations and outreach plans for other observatories and space missions.

## Hubble's History

The arc of Hubble's complicated, high-profile history no doubt contributed to so many people recognizing it as *the* premier astronomical observatory (Zimmerman, 2010). As far back as 1923, rocket pioneers Hermann Oberth, Robert H. Goddard, and Konstantin Tsiolkovsky described how a rocket could launch a telescope into space (Oberth, 1923). In 1946, not long after it became clear that substantial satellites could be launched into orbit with post-World War II rocket technology, astronomer Lyman Spitzer proposed a space observatory, enumerating the assets provided by orbiting above the atmosphere (Spitzer, 1946, p. 546). It was a few decades later that funding and technology actually allowed the launching of telescopes. A few NASA missions such as the Orbiting Astronomical Observatory and the International Ultraviolet Observatory not only proved the concept but also provided groundbreaking science results, though with a relatively narrow range of capabilities.

By 1968, plans were developing to launch a telescope with a 3-m-diameter mirror known then as the Large Space Telescope (LST[1]) (Okolski, 2008). In the next decade, the size of the proposed telescope was reduced to 2.4 m to cut costs but also to make it more practical (Andersen, 2007; Roman, 2014). With funding of $36 million approved by Congress in 1978, LST design could begin seriously with an original target launch date of 1983. In that year, the telescope was named for Edwin Hubble, whose observations confirmed theoretical work by Georges Lamaître that the universe is expanding (NASA, 2006). But many factors, most notably the *Challenger* space shuttle accident, delayed the telescope's launch until 1990.

In a field where the size of the telescope is a defining characteristic, Hubble is a modest-sized research telescope by modern standards. Its size was dictated by the capacity of the space shuttle that launched it and the physical and financial

---

[1] ST, for space telescope, or LST, for large space telescope, were the acronyms used for the project before it was named the Hubble Space Telescope.

constraints of lifting heavy objects into orbit. Nonetheless, Hubble is larger and more powerful than its predecessors in space, and even before its launch in 1990, Hubble was anticipated to provide unprecedented discoveries because of its larger size and extremely diverse capabilities compared with earlier missions. Unfortunately, shortly after launch, it became apparent that there was a major problem. The primary mirror, which collects light and produces the images, was fabricated with a flaw that meant the scientific return was severely compromised. The telescope that was already famous for its promise of discoveries now became infamous as a "techno-turkey" as dubbed by then-Senator Barbara Mikulski, in whose state of Maryland STScI and NASA's Goddard Space Flight Center (GSFC), the overall home of the project, reside (Washington Post, 1993).

The flaw, known as spherical aberration, a minuscule error in the shape of the mirror, turned out to be perhaps the best case of a failure mode. Scientists and engineers were able to characterize the nature of the mirror flaw extremely precisely by analyzing the images of stars. A set of relatively simple corrective optics were designed that could entirely compensate for the misshapen primary mirror. Space Shuttle astronauts installed these optical corrections during a long, dramatic repair mission in late 1993. The resulting images released not long after proved that all the calculations and the risky, dangerous repair mission was a complete success. In the words of Senator Mikulski, "The trouble with Hubble is over!" (C-SPAN, 1994). Once again, the public's attention was drawn to Hubble, this time with a classic redemption story, complete with live video of space-walking astronauts.

## The Importance of Imaging, Dissemination of Information, and Publicity

From the very earliest days of the Hubble Space Telescope, project planners recognized the need to communicate discoveries from the telescope to the public. Aside from the mandate in NASA's charter to inform the public about results from its missions, astronomers wanted the stories to be told.

The document outlining the structure of Hubble's science operations was *Institutional Arrangements for the Space Telescope*, a report of the National Academy of Sciences also known as the Hornig Committee report. It describes the basis for establishing the STScI and includes specific mention of communication and outreach to the public: "... the Institute should be a principal channel for disseminating information regarding the progress and operation of the ST [Space Telescope] and for communicating scientific knowledge derived from the ST to the public" (National Research Council, 1976).

The report also includes mention of image processing capabilities, though primarily regarding data reduction or pipeline processing. This was assumed to take place at GSFC: "The Institute does not need to duplicate the full image-correction capability at the Goddard Space Flight Center (GSFC), which requires a very large computer, nor need it have access to a high-capacity land line" (National Research Council, 1976). In the mid-1970s when the Hornig Report was written, the computing environment was much different than a few decades later when STScI was actually operating. Data processing was much more extensive and less centralized. STScI was established as the center for the project's science operations, including data processing (calibration), archiving, and analysis, in addition to the engineering activities taking place at GSFC.

At the time STScI was established, it included a news office responsible for communicating Hubble's discoveries to the public through news media, working closely with Hubble scientists and in collaboration with the public affairs offices of GSFC and NASA Headquarters. The STScI news office, which eventually became part of STScI's Office of Public Outreach, also had responsibility for producing graphics and images to accompany press releases describing Hubble's discoveries. Press activities began before launch to communicate Hubble's expected results. Writers, graphic artists, and imaging and video producers developed news packages comprising press releases, photos, illustrations, and video which they distributed via the news media and the World Wide Web.

From the beginning of the Hubble mission, NASA's formidable publicity resources helped communicate that Hubble was in fact doing useful science despite operating below its full capability. The message was backed up by the efforts of the science community to develop image processing techniques implemented in software to recover much of the detail in the images lost to the optical aberration. The "deconvolution" techniques improved the images but were not a complete solution to the problem. Nevertheless, the resulting images helped get the message across that the telescope was working better than many people thought. And after the wildly successful servicing mission, STScI and NASA were able to present numerous groundbreaking results and many unprecedented images to communicate the Hubble's achievements.

## Hubble Changed the Public's View of the Universe

The Apollo 11 mission that sent three American astronauts to the lunar surface was extremely dramatic and captured the public's imagination because it was full of firsts; nothing was so dramatic as the first human footsteps on another celestial body. Subsequent space missions, even involving human space flight, have not had quite such an impact, reaching an almost routine level despite the extremely high risks.

Similarly, the Voyager missions to the outer planets were the first to show us images of vastly distant and unusual places up close and from a totally different viewpoint than is possible from Earth. Even in its first few years, however, references to

Hubble in the media exceeded those of the Voyager missions (Christian and Kinney, 2004). Well before launch, the primary message about the Hubble mission was the improvement of science results over those achieved by any ground-based telescopes. So naturally the expectations were high that we would see greatly improved imagery and groundbreaking science. Because of the originally flawed optics, everyone associated with the Hubble project as well as the public in general was disappointed initially by the less-than-perfect results.

Once the optics were repaired, though, the telescope operated as well as or even better than originally expected, and Hubble became the first space science mission to provide many clear, colorful, dramatic images of cosmic landscapes in the deep universe and to capture the imagination of a large fraction of the general public. Comparisons were made with images from Hubble before and after servicing to show the dramatic improvements (see Fig. 4.1A and B). Some of Hubble's subjects had been seen many times in previous images, but Hubble rendered them with much greater detail. Many other subjects were unfamiliar territory—whole new classes of objects largely unfamiliar to most nonastronomers.

Hubble's imagery—indeed most astronomical imagery—does not necessarily include subjects recognizable to most of the general public unfamiliar with astronomy. They are essentially abstract, including forms and other-worldly colors that nevertheless make for compelling, dramatic compositions invoking positive reactions among many viewers. But the images are fully representational: they are real photographs of real subjects, albeit unimaginably distant landscapes.

**(A)**                                    **(B)**

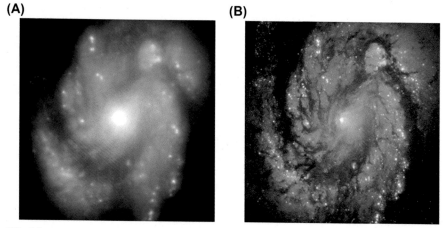

FIG. 4.1

Hubble images of spiral galaxy M100. *NASA and ESA. https://hubblesite.org/contents/ news-releases/1994/news-1994-01.html. (A) Image taken before the 1993 servicing mission with the WF/PC instrument. (B) Image taken in 1993 with the newly installed WFPC2 instrument.*

By understanding some of the "iconography" of the images, most viewers can get a much deeper appreciation of both the visual appearance and intellectual content of the images. Once we recognize what galaxies look like, for example, one can recognize them as the subject of many of the images. And further, once we recognize the varied forms of different classes of galaxies, we can perhaps appreciate the dynamic forces sculpting them. Similarly, we can begin to recognize the components within galaxies, and especially those relatively close, within our Milky Way galaxy. We see nebulae—vast clouds of gas and dust from which stars form and begin their existence. We also recognize different classes of nebulae that are the end stages of stars' existence or the results of supernovae—titanic explosions that entirely disintegrate massive stars and return their material to surrounding space. With Hubble's much higher resolution, similar details can be studied in very distant galaxies. At greater distances, we are seeing these phenomena as they were millions to billions of years ago because of the finite speed of light. Relating an understanding of nearby phenomena with their counterparts in an earlier epoch illuminates the time dimension: we recognize that the cosmos is not static, but constantly changing with the birth and death of stars and galaxies.

It is important to note that the clarity of Hubble's images is largely due to its position orbiting above Earth's atmosphere, resulting in the benefits of avoiding the distortion applied by the atmosphere. And Hubble has its limitations, imposed by the constraints of technology. Earth-bound observatories have much larger telescopes that allow them to collect more light for more efficient observing and potentially better resolution, which can be realized with techniques such as adaptive optics and image deconvolution. Moreover, different space-based detectors can record light beyond Hubble's sensitivity from the near-ultraviolet though the visible into the near-infrared portion of the electromagnetic spectrum. The Chandra X-ray Observatory (Tucker, 2011) extends the view to high-energy cosmic phenomena that produce X-rays. Conversely, the Spitzer Space Telescope extended the view of the universe to the lower energies of infrared light. And even beyond that, radio observatories provide yet a different view of the universe. All research observatories, whether on the ground or in space, collaborate to advance the science by providing data and analysis appropriate to their area of specialization.

Nonetheless, Hubble's imagery raised the "production values"—the technical and aesthetic standards—for astronomical imaging from professional observatories. The Hubble project recognized the need for high-quality, powerful imaging to accompany announcements of discoveries so that the messages were received loud and clear. Nevertheless, there was no conscious intention, nor any need to pursue heroic processing techniques, to achieve such results. The nature of the data available from Hubble accomplished this. Hubble data were the highest quality available to date because of the vastly improved resolution, contrast, and other qualities over previously available astronomical imaging from ground-based telescopes. Careful processing is instrumental in achieving the best results by reproducing all the quality inherent in the data—primarily compositing multiple images into color products.

Indeed, through a large gallery of compelling images, the view of the universe is now largely defined by images from Hubble. For example, many of the entries on astronomy topics or specific astronomical objects in Wikipedia are illustrated by Hubble images because the Hubble image is often the highest-quality image of a given subject. And as shown by media monitoring resources such as Meltwater, other space telescopes have not enjoyed the same level of familiarity among the public despite their contributions to groundbreaking, Nobel-Prize-worthy science and dramatic imaging.

## The Power of Color

Our view of the world is defined by our immediate perception, but the world beyond our immediate view is defined by the images provided by explorers to distant lands. Thanks to today's powerful cameras, our view of distant landscapes looks very different now than in the days of romantic landscape painters or the sharp, black-and-white photography of Ansel Adams.

Similarly, our direct perception of the universe is limited to our view of the night sky, while firmly anchored on the ground. As awesome as that view can be (from a dark, clear site), explorers with telescopes on the ground, and now in space, vastly expanded that view to render images of incomprehensibly distant objects by recording extremely faint light sources. Once we saw only black-and-white photos of familiar subjects seen directly through large telescopes. Now we see extremely detailed views, in bright colors, of not only the old familiar favorites but increasingly unfamiliar, much more distant objects, such as in Fig. 4.2.

**FIG. 4.2**

Hubble image of star-forming nebula Westerlund 2 WFC3 , 2015. *NASA, ESA, A. Nota (ESA/STScI) and the Westerlund 2 Science Team. https://hubblesite.org/contents/news-releases/2015/news-2015-12.html.*

Astronomical photographs have been reproduced in color for some time, but it was at first quite rare because of the complexity of producing color images from the monochrome exposures made with analog photographic technology at state-of-the-art observatories. Color images were produced by Palomar, Kitt Peak, and other world-class observatories, but they were seen mostly in specialty publications. Color "plates" showing many of these iconic images appeared in popular book and college textbooks on astronomy, such as *Exploration of the Universe* by George Abell (1969), a standard text used in first courses for majors in the 1970s (and my first college astronomy text). Even most mass-market print publications such as newspapers and magazines were still in black-and-white halftone reproduction.

The advent of digital photography throughout the imaging industry, including in astronomy, changed that rapidly and dramatically. Among the first color images from earlier space science missions were the exquisite images from the Pioneer and later the Voyager missions to the outer solar system. These revealed landscapes of unfamiliar planets and moons in unprecedented detail and vivid colors and reached wide distribution through (pre-Internet) mass media.

Hubble was part of the digital imaging revolution. The availability of reliable, high-quality, charge-coupled device (CCD) image sensors coincided with the development of Hubble's first camera, the Wide Field and Planetary Camera (WF/PC), among the first use of such devices in research astronomy. Earlier electronic, video-based detectors had been in use in previous space astronomy missions and were used in Hubble's faint object camera (FOC). Digital imaging made the processing and analysis of the data much more rapid and convenient as well as quantitatively more precise and reproducible. The relative convenience of digital compared with darkroom processing made the development of color images from the data much more practical.

An often-asked question is whether the colors in Hubble's images are "real." One interpretation of this question is whether the colors are what we would perceive visually if we could, say, look through the Hubble Space Telescope. The glib, evasive answer is yes and no. But that is because the question is somewhat beside the point and misleading. The real question is whether the colors are artificial, and the answer is no. The colors are a result of the physics of the light sources, not anyone's imagination. False or applied colors are used in astronomy imaging, for example, to colorize a monochrome image. Human perception is more sensitive to variations in hue rather than small changes in brightness or gray value so the color can highlight small variations in structure in the image. Think of weather radar images: the colors are added to monochromatic representations of data to encode the quantitative data in a meaningful way. In the case of radar images, a palette of colors represents varying rain intensity.

But the color composition paradigm combining multiple color bands visualizes the actual colors inherent in the recorded data (Rector et al., 2015). Color results from ratios between the brightness of features in the source at different wavelength bands. Yet, the rendered colors are most often not what we would perceive if we could magically fly close to the sources. For one thing, the light is in general too faint for the human visual system to even perceive color at all. Even if we could get closer to many of the sources observed by powerful telescopes, such as gaseous nebulae, they still would not appear much brighter. The same amount of light we see over a small field of view from very far away is spread over a larger field of view as we get closer. The brightness of stars does depend on the distance though. At the distance to most stars (except the Sun), their disk is not resolved, so the brightness follows the "inverse square" law, which states that the brightness varies as the square of the distance.

Telescopes with a very large aperture, such as Hubble, amplify the light by capturing much more than do our relatively tiny pupils. In fact, the entire imaging technology path, from light captured by the optical path of the telescopes and cameras to the filters and detectors, operates very differently from the human visual system. Filters and processing can be calibrated to result in colors close to be what we would see. But why? The whole point of observing the sources and visualizing the data is to glean information we cannot obtain by using our eyes. Rendering the images in a different, more sophisticated way can reproduce more of the information inherent in the data. Hubble's images are dramatic and powerful not because of the postprocessing but because of the unprecedented quality of the source data from the telescope.

From the very beginning, Hubble mission managers recognized the power of color imaging and its response among the public. The project produced and distributed color photos to the media and the public. The very complete, professional photo lab at STScI produced numerous color photographic prints of images and illustrations accompanying every press release announcing Hubble's discoveries and distributed them to news media (somewhat ironically converting the originally digital images back into analog photographic materials). These were reproduced countless times in print and broadcast media and increasingly in color in publications commonly printed in color. With the Internet and the World Wide Web expanding rapidly in the 1990s just as Hubble was becoming fully active, the images became distributed, both directly from STScI and NASA and via news outlets and other third-party sites, to a wide audience increasingly hungry for more and more colorful, dramatic imagery.

## Hubble Heritage

Carefully produced astronomical images, like those from Hubble and other observatories, highlight the complementarity between the technical, scientific content of the data and the aesthetic value of the images. These are not mutually exclusive; moreover, each aspect heightens the other. An aesthetic appreciation of an image can draw in the viewer and increase a sense of curiosity to learn more of the content. Conversely, understanding the deeper context of an image can expand the appreciation beyond an aesthetic object.

Most of the Hubble images seen by the public were produced to accompany news announcements of major science discoveries. Some of these images were necessarily not the most aesthetically pleasing—after all, much of the groundbreaking science required the telescope to operate at the limits of its capability. Images made with very long exposures of extremely distant and faint sources are close to the technological limits of the instruments. In technical terms, the signal is just above the noise. The images are indistinct with few recognizable features, though deep analysis of the data may yield stunning results. The standard joke is that cutting-edge astronomy is all about fuzzy red dots.

Conversely, many of the observations producing more aesthetically interesting images did not necessarily produce cutting-edge science. Often, the images are of well-known and well-studied targets. Certainly, Hubble's observations have led to a better understanding but may not be in the territory of "discoveries" warranting broad publicity. Nevertheless, the images may be visually compelling, with strong colors and dramatic compositions. If they are of recognizable and well-loved sources, such as the Ring Nebula or the Whirlpool Galaxy, then the public, and especially the scientifically savvy segment, tends to have a stronger affinity for the images.

Nearly a decade into Hubble's history, a few STScI astronomers recognized the sorts of images that would have the greatest public appeal and wanted more of the images to be distributed widely. They proposed a project to work with the rest of the science community to mine the archive to find those data sets that might produce additional beautiful images. STScI management accepted this proposal and provided a small amount of funding, primarily for staff time, to establish the Hubble Heritage Project in 1998, comprising a small team of astronomers and technical experts (http://heritage.stsci.edu/index.html). The overall mission of the Heritage Project was to identify and publicize the aesthetically best images from Hubble while balancing the aesthetic value of the imagery with their science content. Another goal was to develop a definitive gallery of the best images and to foster interactions between the arts and science communities (Noll, 2020).

STScI management acknowledged the benefits of devoting some (albeit small) amount of Hubble's valuable observing time for outreach—that is, to produce images for public consumption rather than purely scientific goals. These observations were used partly to augment existing archival data to produce more complete images: adding a third filter color to produce a more complete color image or adding another field of view to fill in missing pieces of a compelling composition, for example. In some cases, Hubble Heritage was able to image wholly new subjects, which eventually became some of the most compelling Hubble images.

Prompted by a sensitivity not to waste precious science observing time, one guiding principle of the Hubble Heritage project was to produce scientifically useful data sets and not just pretty pictures. Many of the resulting data sets were in fact used in scientific analysis and were cited in technical papers, as documented in the cross-referencing of Hubble observations with published results in the Hubble archive. (See, for example, the first Heritage observing program proposal, http://archive.stsci.edu/proposal_search.php?mission=hst&id=7632.)

The success of Hubble Heritage over many years convinced astronomers at STScI and in the astronomy community more broadly that outreach is an important factor in the public's acceptance of large science initiatives such as Hubble. Many now agree that allocation of a small fraction of the observing time to outreach pays off in the value of the mission to the public. This carries forward as a lesson to subsequent missions such as the James Webb Space Telescope (JWST) and the Nancy Grace Roman Space Telescope (originally known as the Wide Field Infrared Space Telescope (NASA, 2020)), which explicitly include outreach components in their mission plans.

The power and importance of outreach was brought home dramatically after the loss of Space Shuttle *Columbia*. NASA management canceled the subsequent Hubble servicing missions as too great a risk for Shuttle astronauts. But the outcry among the public, the science community, and even the US Congress was considerable. Many of those arguing for reinstating the mission cited the images from Hubble as such a valuable resource that Hubble should be saved (National Research Council, 2005). These arguments proved successful, and a final servicing mission to Hubble launched on May 9, 2009, aboard Space Shuttle Atlantis (Dick, 2014; McBride, 2004).

## Case Studies

A few specific examples show the power and reach of Hubble's results through its history in terms of both the science and the images intended for the media and the public to illustrate announcements of the science results. These examples helped propel Hubble into broad public awareness.

### Comet Shoemaker-Levy 9

The impact of Comet Shoemaker-Levy 9 on Jupiter in 1994 was a singular real-time event that captivated the science community and the public because it was dramatic and nearly unprecedented. Hubble had a "front row seat" to the action and, among all observatories, produced the highest-resolution images of the consequences of the impacts. Fortunately, the impact happened just after the servicing mission to repair Hubble's optical flaw.

This fortuitous confluence of events emphasized Hubble's unique capabilities and kept the observatory in the public eye for quite a while. Perhaps the most significant outcome of witnessing a comet crash into Jupiter was a greater awareness of the dangers to Earth from comets, asteroids, and other rocky fragments populating the solar system. Popular culture responded by producing the films *Armageddon* and *Deep Impact* in the late 1990s (weather.com, 2013). Even the US Congress was prompted to investigate the dangers posed by Near-Earth Objects (Gorman, 2016).

## Eagle Nebula

One of the most widely distributed and recognized Hubble images ever is the Eagle Nebula, also known as Messier 16, or simply M16 in astronomer parlance (see Fig. 4.3). It has also been dubbed the "Pillars of Creation" because stars are forming in pillars of interstellar gas and dust. This target was observed in 1995, and the images were released to the public shortly afterward, instantly achieving unprecedented distribution. Since then, it has achieved something like iconic status, with the image appearing on clothing and jewelry and in other unlikely places; for example, hundreds of items featuring the image are for sale on a curated page on the Etsy website (https://www.etsy.com/market/pillars_of_creation). The image also appears in *Time* magazine's online collection entitled "100 Photos: Most Influential Images of All Time" (Goldberger, Moakley, and Pollack, undated, http://100photos.time.com/).

There are many possible explanations for its popularity, from its aesthetic appearance to the wide distribution possible by then-new Internet technologies to the science story behind the image, many aspects of which were likely unfamiliar to the general public. A particularly curious aspect of the image is how well the form of the pillar structures matches the odd shape of the field of view of the Wide Field and Planetary Camera 2 (WFPC2), the instrument which captured the image. The shape of the WFPC2's field of view results because the camera consists of four components with different optical systems recording adjacent areas of

FIG. 4.3

Hubble image of the Eagle Nebula, M16, WFPC2, 1995. *NASA, ESA, STScI, J. Hester and P. Scowen (Arizona State University). https://hubblesite.org/contents/news-releases/1996/news-1996-01.html.*

the sky, one of which has a smaller field of view. The observers who selected the target were very familiar with both the details of the WFPC2 instrument and the appearance of the Eagle Nebula from ground-based telescope images. They selected this target as a particularly good early test of Hubble's capabilities. Perhaps, however, even they did not imagine how widely the image would be admired.

The image no doubt owes much of its appeal to the more technical aspects of the image: its unprecedented clarity, its unusual color, and the camera's field of view. However, part of it must also be the aesthetics of the image: those pillars look like dramatic, glowing, three-dimensional objects, either geological or biological (Kessler, 2012). In fact, there is an interesting, not entirely coincidental correspondence to more familiar, terrestrial phenomena. The pillars in the Eagle Nebula are the result of energetic starlight and stellar winds eroding away less dense material from vast clouds of gas and dust, sculpting pillar shapes where denser caps shield some of the material. This is analogous to the fantastic spires of rock known as hoodoos— narrow pillar formations with a resistant "cap" protecting softer underlying layers from water and wind erosion—seen in the desert of the southwestern United States such as in Bryce Canyon National Park. Certainly, these are vastly different phenomena in spatial and time scales, but the visual and physical analogies are interesting.

## Hubble Deep Field

The Hubble Deep Field and its successors (collectively, "the Hubble Deep Fields") are likely the most important legacy of Hubble's science results. But it was something of a gamble to which to devote a great deal of precious Hubble observing time, with no guarantee of useful results, not long after the first servicing mission. It was a dramatic exploration of the unknown. As it turned out, the gamble paid off in spades, and many additional observations with later generations of instruments have only increased the scientific return (see Fig. 4.4).

The Hubble Deep Fields' resulting images are not as aesthetically powerful as many other Hubble images. But they are a great example of images that reveal a depth and power when the viewer begins to understand the story an image represents. There is a form of iconography in astronomical images. As the viewer "decodes" the elements of the image, he or she may appreciate the image far beyond the initial aesthetic appearance.

In the Deep Fields, we have a demonstration of the "powers of ten" in a single image. That is, we can appreciate the vast scale of the universe by stepping from our relative neighborhood to the most distant reaches of the cosmos. We are viewing the entire depth of the vast universe, from nearby stars in our own Milky Way Galaxy thousands of light-years away, to galaxies millions of light-years away, to some of the most distant objects every observed billions of light-years away. And the images represent a vast span of time because of the finite speed of light. The light from the most distant objects takes billions of years to get to us, so we see and study those objects in a snapshot of the light from the objects as they were billions of years

FIG. 4.4

The Hubble Deep Field, WFPC2 1995. *NASA, ESA, STScI, R. Williams (STScI), and the Hubble Deep Field Team. https://hubblesite.org/contents/news-releases/1996/news-1996-01.html.*

ago, not as they are now. These images demonstrate that Hubble and other telescopes are time machines as much as they are cameras.

To astronomers, the data comprising the Deep Fields are a treasure trove of information. The Deep Fields shed light on the structure of the earliest galaxies to form after the Big Bang and provide insights on the development of those galaxies from little more than large clusters of stars to the immense conglomerations of stars, gas, dust, planets, and more that we observe in more nearby, contemporary galaxies. In addition, observations of the same fields over many years and over a broad range of wavelengths along with many complementary observations from numerous observatories have provided a much better understanding of the overall structure and evolution of the universe.

Because of these paradigm-shifting science results, the Deep Fields are considered among Hubble's most significant data sets. Numerous scientific papers as well as popular articles highlight these results and the accompanying images have been reproduced repeatedly in news media, books, documentaries, and other outlets. But beyond the scientific importance, the Deep Fields are an example of Hubble results that have inspired creativity in fields traditionally far removed from science and technology. One example is *Deep Field: The Impossible Magnitude of Our Universe*, an orchestral, choral work and a motion picture conceived by the composer Eric Whitacre. It was produced in collaboration with Robert Williams, the principal investigator of the original Hubble Deep Field observations, with technical input from the Space Telescope Science Institute (https://deepfieldfilm.com/).

## Making Hubble Images Increasingly Accessible

The concurrent deployment and operation of Hubble with the explosive rise of information and communication technology, and the Internet in particular, turned out to be a perfect storm for Hubble's results to be disseminated to an extremely wide, diverse, international audience. Compelling events surrounding the history of the telescope—from its scientific promise to a disastrous error in optical fabrication and subsequent risky astronaut repair missions to a steady stream of groundbreaking discoveries and compelling images—served to amplify the high profile of the mission in the public's consciousness. Huge numbers of people had instant access to the results and the images via the Internet. The World Wide Web and later social media were just getting started as traditional media were waning as the primary source of news and information (Christian, 2004).

In keeping with the spirit of the Hubble Heritage Program, STScI continues to release images with largely aesthetic appeal rather than for a specific scientific purpose. In particular, major anniversaries of Hubble's launch are commemorated with a new release, usually with wide distribution in traditional and social media. The most recent example to date was an image of the star-forming nebula NGC 2014 and NGC 2020 produced in April 2020 (Fig. 4.5). Reactions to this release were widespread, with estimates of potential readers soon after the release approaching one billion (Villard, 2020).

Press officers at STScI and NASA centers lamented the demise of the dedicated science reporters at major news media and even numerous smaller-market local newspapers and broadcast outlets. Many of these accomplished reporters were expert at translating the sometimes-esoteric subjects of news releases from NASA and science centers into stories digestible by average readers. (A few of the press

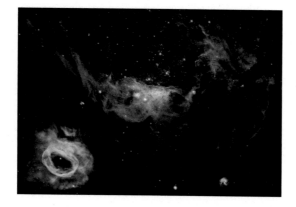

FIG. 4.5

Hubble "Cosmic Reef" image of the giant red nebula (NGC 2014) and the smaller blue nebula (NGC 2020) in the Large Magellanic Cloud, WFC3, 2020. *NASA, ESA, and STScI. https://hubblesite.org/contents/news-releases/2020/news-2020-16.*

officers, particularly those connected with NASA's Great Observatories, were expert at crafting news releases that explained the science in a way that was understandable, at least to these reporters.) Nevertheless, the same press officers certainly enjoyed the broad distribution of the results from Hubble and its sister missions through the more direct means of the Internet. And the press offices became more adept at crafting the stories and the accompanying visuals for the new medium of web sites.

While the influence of traditional news outlets such as print newspapers and nightly TV news has declined in favor of widespread digital access, additional new technologies and resources have also provided expanded access to images and science results from Hubble and other observatories to formerly underserved communities. Perhaps most notably, Noreen Grice pioneered the production of books with tactile representation of astronomical images and braille text (Grice, 2002). Others have produced 3D-printed versions of images with varying textures that represent features in the images (Kramer, 2014; NASA, 2014). Future observatories such as JWST have already incorporated efforts to broaden their reach to underserved communities into their outreach programs by leveraging new technologies (Meinke, 2017).

## Conclusion

The Hubble Space Telescope is likely the most famous observatory in history. It is also one of the most scientifically productive experiments ever. Hubble's scientific power is a testament to the foresight of many scientists and engineers who designed the mission as well as the army of dedicated technical people who built it and continued to keep it operating.

Hubble's prominence in the public's consciousness is due to the remarkable, groundbreaking science results over many years of operation. Hubble's complex, dramatic history must also play a part in its high profile: a near disastrous failure and remarkable redemption, made possible by risky human space flight, periodically renewed with repeated servicing missions. But just as important must be a growing gallery of dramatic, powerful images that have brought unimaginably distant spacescapes to our printed pages, computer screens, and mobile devices. Hubble has done nothing less than reimagine our collective view of the universe.

Future observatories such as JWST have already taken note of the broad public appeal of space imagery and are incorporating efforts to engage the public into their outreach plans. Indeed, publicly funded missions such as NASA's rely on the support of the public to encourage policy-makers to continue funding these large projects. A measure that has been successful for the Hubble mission is to devote a small amount of resources to produce images with strong aesthetic appeal to complement groundbreaking science results.

# References

Abell, G. O. (1969). *Exploration of the universe*. New York: Holt, Rinehart and Winston.

Andersen, G. (2007). *The telescope: Its history, technology, and future*. Princeton University Press.

As one example, the first heritage observing program: HST heritage program proposal 7632.Retrieved from http://archive.stsci.edu/proposal_search.php?mission=hst&id=7632.

Christian, C. A. (2004). The public impact of the hubble space telescope: A case study. In A. Heck (Ed.), *Organizations and strategies in astronomy. Astrophysics and space science library* (Vol. 310). Dordrecht: Springer.

Christian, C. A., & Kinney, A. *The public impact of hubble space telescope*. Retrieved from https://amazing-space.stsci.edu/eds/epo/monographs/PIHSTMono_102.pdf.

*Deep Field Movie*. Retrieved from https://deepfieldfilm.com/.

Dick, S. J. (2014). The decision to cancel the Hubble space telescope servicing mission 4 (and its reversal). In R. D. Launius, & D. H. DeVorkin (Eds.), *Hubble's legacy*. Smithsonian Institution Scholarly Press.

Goldberger, B., Moakley, P., & Pollack, K. (undated). *Most influential images of all time*. 100 Photos. Retrieved from http://100photos.time.com/.

Grice, N. (2002). *Touch the universe: A NASA braille book of astronomy*. Washington, DC: Joseph Henry Press. https://doi.org/10.17226/10307. Retrieved from https://hubblesite.org/contents/news-releases/2008/news-2008-05.html.

HDF. Retrieved from https://hubblesite.org/contents/news-releases/1996/news-1996-01.html.

*Hubble heritage*. Retrieved from http://heritage.stsci.edu/.

Kessler, E. A. (2012). *Picturing the cosmos: Hubble space telescope images and the astronomical sublime*. University of Minnesota Press.

Kramer. (2014). Retrieved from https://www.space.com/24233-3d-printed-hubble-photos-blind-aas223.html.

McBride, K. (January 29, 2004). Hubble, the beloved: Decision to stop maintenance of telescope generates outpouring. *Washington Post*.

Meinke, B. et al. Emerging technologies in outreach for JWST.2017 Retrieved from https://ui.adsabs.harvard.edu/abs/2017DPS....4910104M/abstract.

Mikulski, S. B. (January 13, 1994). The trouble with hubble is over. In *Hubble space telescope NASA news conference, C-SPAN*. Retrieved from https://www.c-span.org/video/?53735-1/hubble-space-telescope.

NASA. (2014). Retrieved from https://www.nasa.gov/content/goddard/hubble-images-become-tactile-3-d-experience-for-the-blind.

NASA. (2020). Retrieved from https://www.nasa.gov/press-release/nasa-telescope-named-for-mother-of-hubble-nancy-grace-roman.

National Research Council. (July 19–30, 1976). *Institutional arrangements for the space telescope: Report of a study at woods hole, Massachusetts*. Washington, DC: The National Academy Press.

Noll, K. (April 28, 2020). *Private communication*.

Oberth, H. (1923). *Die Rakete zu den Planetenräumen. R. Oldenbourg-Verlay* (p. 85).

Okolski, G. (2008). *A chronology of the hubble space telescope*. NASA. Retrieved April 26, 2020.

*Pillars of creation*. Retrieved from https://hubblesite.org/contents/news-releases/1995/news-1995-44.html.

Rector, T. A., Arcand, K., & Watzke, M. (2015). *Coloring the universe: An insider's look at making spectacular images of space.* University of Alaska.

Retrieved from https://www.etsy.com/market/pillars_of_creation.

Retrieved from https://www.washingtonpost.com/archive/politics/1993/11/28/mission-has-big-risks-for-nasa/91f3642e-d087-448c-8a20-4eb30c4f538b/.

Roman, N. G. (2014). Conceiving of the hubble space telescope. In R. D. Launius, & D. H. DeVorkin (Eds.), *Hubble's legacy.* Smithsonian Institution Scholarly Press.

Spitzer, L., Jr. (1946). *Report to project rand: Astronomical advantages of an extra-terrestrial observatory.* (Reprinted in NASA SP-2001-4407: Exploring the unknown, chapter 3, document III-1).

STScI-PRC94-01 M100. Retrieved from https://hubblesite.org/contents/news-releases/1994/news-1994-01.html.

*The weather channel.*(2013). Retrieved from https://weather.com/science/space/news/comet-hits-jupiter-20130220.

Tucker, A. (2011). Retrieved from https://www.smithsonianmag.com/science-nature/brilliant-space-photos-from-chandra-and-spitzer-60596/#vb0HyvTaV2DvLxof.99.

Villard, R. (April 29, 2020). *Private communication.*

*Westerlund 2.*(2009). Retrieved from https://hubblesite.org/contents/news-releases/2015/news-2015-12.html.

Zimmerman, R. F. (2010). *The universe in a mirror: The saga of the Hubble space telescope and the visionaries who built it.* Princeton University Press. ISBN 978-0-691-14635-5.

https://www.anseladams.com/. (Accessed 8 March 2021).

## Further Reading

*Harvard-smithsonian.*(2013). Retrieved from http://chandra.harvard.edu/photo/false_color.html.

*The path to hubble space telescope.* (2006). NASA. Retrieved from https://web.archive.org/web/20080524211736/http://hubble.nasa.gov/overview/conception-part1.php.

Gainor, C. (2021). *Not Yet Imagined.* NASA. Accessed 9 March 2021.

# Enabling Learning and Empowering Workforce Preparation Using Space-Based Authentic Experiences in Science, Technology, Engineering, and Mathematics

Sheri Klug Boonstra, Don Boonstra

## Introduction

Space exploration is one of those engaging and interesting topics that catalyzes the imagination and can crystallize the vision of the future in a way that few others do. The inhabitants of our planet have been boldly exploring our solar system and beyond robotically and generating amazing discoveries and images that engage and inspire the world. In addition to the robotic endeavors, serious planning is underway to send humans to live sustainably beyond Earth and become a multiplanet species. These efforts have set in motion a whole range of dreams of what is possible and the necessity of gaining the skills to make it happen.

Providing learners with authentic STEM opportunities that will engage them in first-person experiences that relate directly to their much-imagined futures will help them to appreciate and value the innovation, knowledge, and technology it took to, as NASA's Jet Propulsion Laboratory says, "Dare Mighty Things." Using space exploration as a theme and wrapping learning within that relevant and exciting framework can be highly successful to inspire learning!

In the continuum of formal and informal science, technology, engineering, and mathematics (STEM) learning that students encounter from elementary school through their terminal degree, how can formal and informal educators be more effective and intentional in helping them prepare for their future careers? The world is changing with ever-increasing speed in terms of technology and complexity. Sadly, the textbook curriculum from which many educators instruct their students can be woefully static and outdated. Working from noncurrent curriculum leaves their students at a disadvantage in terms of preparation and competitiveness for those future jobs.

Space Science and Public Engagement. https://doi.org/10.1016/B978-0-12-817390-9.00003-8

Numerous presidential and national councils and committees over the past two decades have expressed concern with the need for students' STEM learning to stay in step with our fast-changing world and have encouraged educators to embrace approaches that involve active, engaged learning in which students interact with authentic STEM data in an authentic problem-solving context to derive meaning and construct understanding of the world (President's Council of Advisors on Science and Technology, 2010; Committee on STEM Education National Science and Technology Council, 2013; National Research Council (NRC), 2012; National Research Council, 2005, 2007). According to the US Department of Education's *STEM 2026: A Vision for Innovation in STEM Education*:

> *The complexities of today's world require all people to be equipped with a new set of core knowledge and skills to solve difficult problems, gather and evaluate evidence, and make sense of information they receive from varied print and, increasingly, digital media. The learning and doing of STEM helps to develop these skills and prepare students for a workforce where success results not just from what one knows, but what one is able to do with that knowledge (emphasis added) (Tanenbaum, 2016).*

In other words, these authentic STEM experiences need to be relevant and engaging to the students' world, flexible as to be inclusive of all learners, and challenging in terms of giving students the skill sets and practice toward becoming the problem-solvers and innovators that the future workforce will demand.

In this chapter, we will outline the critical steps necessary to building an authentic space-based STEM experience and emphasize the beneficial preparation that these experiences will provide in shaping and adding needed practice for learners as they move forward in their career pathways. We each have more than 20 years of experience working in NASA missions, creating and directing NASA education programs. In addition, we have over 55 combined years teaching in precollege and higher education classrooms. This chapter will draw on that deep experience and lessons learned to explain the authentic STEM learning environment, supply background, and the application of the framework of standards, provide a guide to best practices and lessons-learned, give examples of exemplary projects, and emphasize the importance of scholarly evaluation. Finally, we will show the need to create or identify "next step" authentic STEM connections for students to keep practicing and acquiring relevant knowledge and skills moving toward their career pathway.

## Building the Case for Authentic STEM Experiences in Education

Research shows that students who are introduced to STEM topics by being immersed within a project that utilizes authentic STEM processes can experience transformational aspects in terms of how they learn (National Research Council, 2007; Tanenbaum, 2016). Taking into account the most recent research about what *has* and *has not* worked in education, in 2012 the *Framework for K-12 Science Education (Framework)* (National Research Council, 2012) and the *Next Generation Science Standards (NGSS)* (NGSS Lead States, 2013) that evolved from the *Framework* provided a new powerful conceptual change in educating students in science and engineering. Notably, the *Framework/NGSS* focuses on what students do with information, not on what they do to obtain information, while also more tightly integrating the content of science and the processes of science. The *Framework/NGSS* calls for learning experiences in various scientific and engineering disciplines to engage students not only though exposure to the content of the disciplines but also in the practices of science and engineering—including asking questions and defining problems, developing and using models and computational thinking, planning and carrying out investigations, analyzing and interpreting data, building arguments from evidence, and communicating results—as well as concepts regularly used by scientists and engineers including cause and effect, pattern recognition, systems thinking, scale and proportion, energy and matter cycles, structure and function, and stability and change.

Authentic STEM experiences can, if structured properly, address these aims. In doing so, educators can create an environment in which students are active participants in learning. In this environment, students are the essential workers in the educational process. They construct, discover, and develop central concepts. They read, write, talk, think, pose questions, and solve problems. They observe and manipulate aspects of their environment and confront resulting problems. They take risks. They exhibit the ability to learn how to learn. They demonstrate understanding of the central concepts and competence with the essential skills in individual and group problem-solving. Students exhibit a willingness to accept different kinds of solutions to the same problem and to work with other students outside of class.

Indeed, authentic STEM experiences providing learning experiences that are not cookbook, one-right-answer-for-the-entire-class will provide much needed relevancy and transform the learning environment to one in which students develop ownership in their learning. Students can gain understandings and obtain practice in critical workforce skills as identified by The Partnership for 21st Century Learning's *Framework for 21st Century Learning* (Partnership for 21st Century Learning, n.d) such as creativity and innovation; critical thinking; collaboration flexibility and adaptability; initiative and self-direction; leadership and responsibility; and information, media, and communication technology literacy. Such skills are essential in many work environments and no less so in the STEM workforce.

Furthermore, authentic STEM experiences can help all students gain fundamental understandings about the nature of science and its value to society, thus contributing to an increase in science literacy in our nation. As a 2010 report of the President's Council of Advisors on Science and Technology (2010) indicated: "In the 21st century, the country's need for a world-leading STEM workforce and a scientifically, mathematically, and technologically literate populace has become even greater, and it will continue to grow — particularly as other nations continue to make rapid advances in science and technology." In the words of former US President Barack Obama, "We must educate our children to compete in an age where knowledge is capital, and the marketplace is global"(WestPoint, 2010).

## Using the Excitement and Relevance of Space Exploration to Promote Learning and to Help Prepare Our Future Workforce

A research team gathers excitedly as their image arrives from a camera on board a NASA spacecraft in orbit around Mars. They pour over the image they targeted to answer a scientific question about the Red Planet. Working together, the team measures, analyzes, and discusses relations among surface features as they seek to answer their research question. This research is not, however, occurring in a NASA laboratory but in a classroom. The researchers, in this case, are in a 7th-grade science classroom in a small rural town in northern California (Klug Boonstra & Christensen, 2013).

These students are not using worksheets to learn their required science content. Instead, they *are* the scientists, using authentic data from NASA spacecraft instruments that are in orbit around Mars to investigate their own research question. In this particular case, they are investigating the flanks of martian volcanoes to look for evidence of skylights (cave openings). The students are engaged in the Mars Student Imaging Project (MSIP) and have spent 30+ hours both in and out of their class looking at hundreds of images of the martian surface to drive this project forward. During this process, they are learning about remote sensing and the instruments that return planetary data. They are using the same technology-driven planetary science tools that professionals are using. They are designing their investigation based on their interests and curiosity about the processes of planetary surface formation, and, just like NASA does, they are working in a team to conduct their research.

Their diligence pays off! They have located a new cave on Mars (see Fig. 5.1), and the discovery goes viral: *"7<sup>th</sup> Graders Discover Cave on Mars!"* (Burnham, 2010).

Indeed, the topic of space has been an enduring way to engage humanity across the ages. Interest in space crosses cultures and geographic lines, inspiring young and old alike. Space sparks human curiosity with each new image of other worlds, whether it be from a fleet of spacecraft exploring our own solar system or from telescopes looking across galaxies and discovering stars with their own planet systems. We want to understand our place in the cosmos and answer the ageless question: "Are we alone?"

With more nations now participating in space exploration, the idea is more conceivable than ever that humans will become a multiplanet species. Living and working *off-planet* presents great challenges with many yet unanswered questions. At the same time, with each successive generation, technology is enabling a

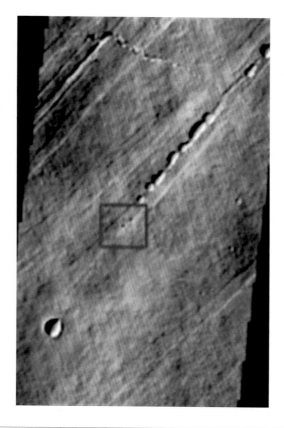

FIG. 5.1

MSIP cave image, THEMIS camera visible image #V3687000. *HiRISE Camera/UofA/ NASA/Caltech.*

different, more global view of Earth—one that sees our planet as a system with inter-connected, fragile complexities that breach geographic, political, and cultural boundaries and has implications for our entire species. If humanity is to successfully address the pressing ecological, social, and political problems on our planet and perhaps reach the desired destinations that lay beyond Earth, it will be necessary to create and adopt new ways to prepare our workforce that will empower them to embrace to new iterative, adaptive ways of applying innovative thinking and problem-solving.

Creating authentic STEM experiences using space exploration as a theme for engagement can achieve these ends. Tying these experiences to relevant current and future exploration challenges not only engages learners in problem-solving as an activity but also gives them experience in thinking about things that have not been done before, such as how to live on another planet. Authentic space-based STEM projects can provide the proving ground and practice for learners as they ready themselves for their future career, whether space-related or otherwise.

## Integrating Space Science into Formal and Informal Learning: The Case of the Mars Student Involvement Project

NASA is in the business of discovery, adhering to a set of mission goals identified by the space science community and approved by political stakeholders. K-12 STEM educational institutions are in the business of preparing students to be science and technology literate and stimulating those who will be the next generation of scientists, technologists, engineers, and mathematicians, all while being held accountable to learning standards and curriculum goals. How do the two come together for the betterment of both?

The answer lies in the fact that both benefit immensely from the study of natural events. Developing spacecraft and research programs to examine the structures, features, compositions, and changes taking place over time on planets and stars is a key component of NASA's charge as the US civil space agency. Meanwhile, authentic STEM experiences inviting students to observe natural events in a variety of disciplines allow them to employ the practices that scientists use in their investigations and theory development and that engineers use to design and construct systems. Carefully chosen natural events that allow students to deeply engage with a discipline and the concepts outlined in the *Framework/NGSS* bridge the obvious chasm between the NASA mission requirements and K-12 classroom requirements. By coupling experienced education professionals who are competent in designing authentic STEM learning experiences with subject matter experts in science and engineering to ensure accuracy of content, the investigation of natural events using data from NASA research domains can be a strong enabler to bring NASA-based authentic STEM learning into the classroom.

NASA's Mars Student Imaging Project (MSIP) is one example of how an authentic science experience has been successfully implemented with over 80,000 students across the United States. MSIP is a nationally recognized, award winning (Klug Boonstra & Christensen, 2013), authentic project-based learning, and student-centered education project. Students learn how STEM works by engaging in authentic research using data from NASA's Mars Odyssey spacecraft, constructing an understanding of core ideas in science and engineering and the process of science and exploration by *being* scientists.

Created to be an immersive and transformational way for students in grade 5 through early college to engage in scientific process and practices, MSIP has been designed to *not be* a "one-size-fits-all" experience. The curriculum has detailed instruction for the instructor to facilitate the project without a lot of expert knowledge in space or, for that matter, science. This allows all instructors to have enough confidence to implement this project successfully. MSIP has also been formulated to be used easily in both formal and informal education settings. The MSIP curriculum is strongly aligned to the *Framework for K-12 Science Education, NGSS*, and the *Framework for 21st Century Learning*. It enhances the teaching of traditional courses, such as physical science, Earth science, chemistry, and life science/biology—an important point considering that if the materials being covered in instructors' classrooms are not strongly aligned to the standards they must cover, then the adoptability and defensibility of the use of space-based curriculum is at high risk.

There were originally three MSIP formats that an instructor could use with students: the On-site at Arizona State University (ASU) option, the Independent Research option, and the Distance Learning option. The On-site option, where classrooms of students did a portion of their MSIP project at ASU, was dropped after a period of time due to cost (time and salary of facilitators) and the scalability (only a limited number of students could use this option, mostly in the ASU area). All MSIP materials are free and can be obtained on the ASU Mars Education website (http://marsed.asu.edu).

The Independent Research option works well if an instructor wants to create her/his own research structure and/or has limited time and cannot accomplish the entire recommended MSIP schedule. Instructors can use MSIP materials any way that makes sense for their class and time allotment. The Independent Research option students would then follow a plan developed by their instructor to fit within their time frame and their instructor's goals. Students have access to the publicly available large database of images already acquired by the Thermal Emission Imaging System (THEMIS, the primary camera onboard the Odyssey spacecraft) or any other data sets already in the Java Mission planning and Analysis for Remote Sensing (JMARS) repository of data from Mars and other planets and small bodies across the solar system. The Independent Research option is limiting in that it does not offer interaction with the MSIP team or any Mars scientists. This option works well for international teams of students that are interested in participating in MSIP.

The third MSIP option (and the most popular) is the Distance Learning option. If an instructor is interested in involving MSIP staff and Mars community scientists in the research project for support and feedback, MSIP offers this format. The basic process for the Distance Learning option first involves students building their understanding of the surface features of Mars (such as craters, volcanoes, gullies, canyons, lava tubes) and the types of images available and how to take data from them through an online lesson. Next, student teams develop a researchable question and hypothesis as well as a research plan to answer their question. Student teams then present their research plan to an MSIP staff member through a virtual meeting platform using a provided MSIP proposal template. The first virtual MSIP project meeting with the MSIP staff member helps to clarify and refine the student question and offer guidance to help focus the student teams in their research. Students teams can either target an image using the THEMIS camera (if available), or they can opt to use the large database of images already acquired by the THEMIS camera or any other data sets in the JMARS database. Lastly, student teams execute their research plan: they collect and analyze the data required to address the hypothesis, agree upon a conclusion to support or refute the hypothesis, develop a presentation of final results (template provided by MSIP), and present the findings to a Mars scientist through a virtual meeting platform. Because MSIP is supported by NASA through US taxpayer funding, the Distance Learning option is only available to student teams located in US schools and international teams formally partnered with US schools.

According to data collected from the educators facilitating this project, MSIP takes about 30 hours to fully implement. When teachers were shown how to implement an authentic STEM lesson and what the return on investment would be for their students—who would be learning how science and engineering work by doing their own unique research instead of "cookbook" labs—teachers reacted favorably to making the time to implement MSIP.

Our research shows that once students have been given the knowledge, skills, and tools to explore space, they become empowered and do not want to learn from worksheets any longer. Because they "owned" their learning, we have seen many teams keep going beyond MSIP to study other bodies within the solar system. We have seen former MSIP students and teams presenting their original research at professional space technical conferences—not in a section dedicated to student research or education and outreach but in a professional, technical session. We have been notified of high-school students who were published as contributing authors in technical proceedings because of their research which had its start in MSIP. We have learned of students getting internships at NASA because they knew how to use professional tools to conduct research and had the attention to the methodology it requires. Finally, an MSIP student was accepted as a middle schooler to present an analysis of a potential landing site on Mars at a workshop of the professional Mars community. These students, once given the opportunity to learn and contribute to something real in the context of an authentic learning experience, have been transformed in the knowledge that they *can* contribute to the world! They have been unleashed and have no desire go back to the worksheet way of learning!

## Strengthening the Workforce Pipeline: Identifying and Connecting Students to the "Next Steps" in Authentic STEM Beyond K-12

As students move beyond K-12 and into higher education, many times the more flexible K-12 ways of learning yield to more traditional forms of lecture classes, labs, and structured ways of instruction. In the National Academies' *Rising Above the Gathering Storm* report (National Academy of Sciences, National Academy of Engineering, and Institute of Medicine, 2007), several key points were called out in helping to provide important, inclusive ways for postsecondary students to prepare for the workforce:

- *Peer support:* Giving students opportunities for interaction that build support across cohorts and promote allegiance to an institution, discipline, and profession.
- *Enriched research experience:* Offering beyond-the-classroom, hands-on opportunities and summer internships that connect to the world of work.
- *Bridge to the next level:* Fostering institutional relationships to show students and faculty the pathways to career development.

In reimagining how students can more fully engage in their higher education preparation toward their careers, there need to be ways for students to apply what they are learning. Having seen the transformational impact using authentic STEM has on students for nearly two decades in K-12, we believe it is imperative that there be avenues of "next steps" for students who have learned through authentic STEM experiences to apply their higher education discipline learning to real-world situations. The use of problem-based learning that is directly mapped to the needs of STEM higher education is becoming an emerging trend. To produce the next-generation workforce who will be ready to be the problem-solvers, innovators, and explorers of the unknown, there must be an intentionality of allowing students to tackle real-world challenges prior to entering the workforce. As programs like the ones we describe below publish their findings, more adopters will likely see the advantages of moving toward this type of instruction.

We applied this logic when creating a NASA-based science and engineering course, as well as a follow-on learning opportunity tied to a NASA mission, to allow students at an institution of higher education to experience authentic STEM learning. Acknowledging that most institutions of higher education do very well at teaching students about their discipline content, we chose a different niche to fill in terms of student preparation. From our observations, higher education students did not often get *applied practice* within their STEM disciplines that would help them see the relevancy of their discipline instruction and apply this knowledge to unsolved challenges. Also, students rarely got the chance to work on challenges that were cross-disciplinary and reflect more realistically the actual workforce environment.

## The ASU Space Works Project

In the spring of 2018, an authentic STEM project, the ASU Space Works 1 Project, was introduced as a three-credit course for undergraduate science and engineering majors at Arizona State University. This introductory workforce development course takes students through a series of cross-curricular trainings which requires them to (1) get qualified in the student machine shop (See Fig. 5.2) to better understand and apply the fundamentals of fabrication; (2) learn engineering drawing and computer-aided design (CAD) software (SolidWorks) to better understand the NASA mission-based, technical design process and modeling; (3) become knowledgeable about NASA and industry standard protocols and practices such as schedules, budgets, proposals, and other processes of NASA and the space ecosystem; (4) establish a high level of communication and be inclusive to tap all the talent within the student team to move their space-focused design project forward; (5) heighten the awareness about the need for the team to work across disciplines to understand both science and engineering goals in relation to the schedule, budget, and deliverables; and (6) work on a student team to create a mission concept within an exploration scenario and deliver their final product.

**FIG. 5.2**

STEM Space Works science and engineering students get qualified in the ASU machine shop to learn how to fabricate mission elements. *S. Klug Boonstra.*

The team is assessed for the course through their deliverables for the Space Works 1 course which include a NASA-standard midterm Preliminary Design Review (PDR) package and a team presentation on their final project work. A review board question-and-answer session follows their presentation facilitated by engineers and scientists who are currently developing NASA mission flight hardware at ASU, and a final End Item Data Package report that covers the evolution and results of the entire project is submitted at the end of the course.

Students are assigned to work in randomized teams at the beginning of the semester to design their concept, fabricate it to the provided mission constraints, test the model (See Figs. 5.3 and 5.4), iterate the model, and test again. Randomized teams are used to better align with what the students will encounter in the workforce. The project teams are chosen by Space Works staff with each team chosen by considering a variety of factors: class standing (e.g., freshmen through seniors, with the more advanced students placed on the teams first as an anchor of skill and knowledge) and diversity of gender, discipline, and culture.

**FIG. 5.3**

Space Works 1 students culminate their semester with a drop test of their mission concept. *Charlie Leight/ASU Now.*

**FIG. 5.4**

Professor and NASA instrument principal investigator Dr. Phil Christensen checks on a student's failed module drop. *Charlie Leight/ASU Now.*

The Space Works students have a dedicated 3000-square-foot maker space that provides them with tools, 3D printers, computers with CAD software, and work and collaboration space so their teams can optimize their designs and projects. Students are mentored by the Space Works staff scientists and engineers and provided instruction by the mission engineers building a variety of NASA solar system flight instruments at ASU within Dr. Phil Christensen's space instrument development group. The initial Space Works course filled to capacity at 60 students and was an overwhelming success in terms of student interest, participation, and achievement of project goals.

Knowing and understanding the necessity and benefit of creating follow-on pathways to more complex learning, a second Space Works course was created for the fall of 2018 and had 40 of the 60 original Space Works 1 students register. Based on our evaluation and follow-up, attrition was attributed to seniors graduating and course schedule conflicts.

The Space Works 2 curriculum focuses on modeling, building, and testing. Students learn an additional industry-standard CAD—Siemens NX software—to build the models of their projects. They learn to work with electronic components within the context of mission design, and their teams were able to construct an electronics project that had to perform within mission constraints when tested in space conditions within Dr. Christensen's thermal vacuum chamber. Randomized student teams are again assigned, and student teams go through routine NASA mission procedures such as creating schedules, budgets, and planning their procurement processes (as they learn in Space Works 1) for the parts needed for their uniquely designed projects. Before getting permission to build their project design, teams deliver a PDR package to the engineering review board for approval. Once the approval is granted, teams proceed to their instrument build and finally test their concepts in space-like (vacuum) conditions. When failures occur, teams have to go before a failure review board to describe the reasons for their failure and what the mitigation will be, rework their instrument, and test again.

In the spring of 2019, Space Works 1 was repeated (again filling to capacity), and Space Works 3 was launched concurrently. Thirty students enrolled in Space Works 3, with some students returning to the course having had to skip Space Works 2 because of schedule conflicts. The Space Works 3 curriculum incorporates structural modeling, again using NX CAD software and additive manufacturing (3D printing). The first 6 weeks of the course focuses on learning structural modeling and testing techniques and includes a vibration test of a professionally fabricated spacecraft instrument model. After this portion of the training, the students form their own teams and design their own experiments applying all the lessons learned in the previous Space Works courses, following the same mission procedures and processes. The student teams can optimize their own designs within a mission concept from a notional Announcement of Opportunity (call for mission proposals) presented to them by Space Works staff and chosen to address near-term needs that NASA would likely be soliciting. This approach gives these soon-to-be early career professionals practice in a relevant area they are likely to encounter early on in the workforce.

Through our efforts of seeing a gap in workforce preparation and filling the gap with cross-discipline, authentic STEM, multisemester courses, the ASU Space Works Project courses have now become foundational courses at ASU, as a joint venture of the School of Earth and Space Exploration (SESE) and the Fulton School of Engineering. The Space Works courses are listed under SESE (which helps support Space Works instructor salary costs). ASU's Fulton College of Engineering considers Space Works courses, which are cross-listed in engineering as technical electives, adequately rigorous such that ASU engineering students can receive engineering credit toward their graduation. Both schools see their discipline instruction being applied in a challenging, real-world way to help their graduates be more practiced in mission-based projects as they enter the workforce.

### NASA's Lucy Student Pipeline Accelerator and Competency Enabler Program: Developing NASA's Diverse Workforce at a National Scale

With the success and lessons learned from ASU's Space Works Program, another opportunity arose within NASA's Lucy Mission to the Trojan asteroids. Funding was available for a Lucy mission student collaboration to expand the Space Works model to a national level. The Lucy Student Pipeline Accelerator and Competency Enabler (L'SPACE) program was initiated in July 2018 with the target program rollout for the fall of 2018. The program was designed by taking elements of the successful Space Works Academy program and applying them nationally through a virtual platform and delivering them weekly through 12-week sessions. These sessions are offered three times a year—spring, summer, and fall—and will continue until October of 2021 when the Lucy mission is planned to launch. The L'SPACE staff are the same core staff that run the Space Works program. In addition, the Lucy mission team has assigned the scientist who handles the mission's outreach portfolio and who is located in Colorado to be part of the L'SPACE leadership team. The audience for the L'SPACE Academy is the same as Space Works: undergraduate STEM students attending US colleges and universities. However, since our program funder, NASA, is interested in diversifying its workforce, there is a special emphasis in the recruiting of underserved students (minorities, female, rural, urban) as well. Community college students, who come more often from underserved populations than students in 4-year institutions and are often the first in their families to attend college (American Association for Community Colleges, n.d), are encouraged to participate, even if they have not formally declared a major but are interested in STEM careers.

For the pilot session of L'SPACE Academy 1, no announcement was sent out via NASA for recruitment of students so that the program staff could work with a smaller set of participants to validate the methodology of the program design; the opportunity was posted on the ASU L'SPACE website and the Lucy mission homepage only. A total of 381 students still found the announcement and applied! The virtual platform chosen for L'SPACE (Zoom) has the capacity to support up to 500 students and L'SPACE staff. The L'SPACE leadership team made the decision to go ahead and accept all 381 students, and the program was launched. Out of the 381-student inaugural class, 351 students (92%) who participated in teams finished the full 12-week program and turned in their deliverable of a mission-related PDR document. Teams were grouped by the time zone in which they were located; in our pilot, we worked across six time zones.

We followed up with all students who withdrew from the program (8%) to better understand the reasons and circumstances of their withdrawal. These data were used to determine if there were strategies that could be employed to help the most vulnerable students participate in L'SPACE in the future. Most of the students who had withdrawn had schedule conflicts, as students were required to attend all the weekly sessions synchronously. The students who withdrew recognized the need to work with their teams to complete the project tasks and felt they would not be able to help their teams at the level necessary. They did not want to let their teams down. Most students opted to rejoin the L'SPACE Academy in the next semester. We are keeping data on all of the students who dropped out to see if they reengage with the program.

The L'SPACE sessions are interactive and give students access to NASA and mission subject matter experts, who help guide and inform the teams in their final projects. If they need to miss a session, students can inform the L'SPACE program ahead of time and receive an excused absence. If excused, the students need to watch the missed session and send the L'SPACE leadership team a synopsis of what they learned and personally took away from the session. The sessions are recorded and distributed to all the students afterward to allow the students to revisit what was presented. To prevent students from not attending synchronously and just expecting to get the session recording anyway, attendance is taken for each session, and students receive a notification if they miss a session. Students must attend a minimum of 10 sessions to receive a program certificate of participation.

Special consideration is given for students who have to work to cover their college expenses. Students are encouraged to access L'SPACE staff for questions or mentoring through the program website to ensure they understand all content. Students also have access to L'SPACE announcements through the L'SPACE website forums and work with their teams on an industry-standard team communication tool (Trello). All students completing L'SPACE Academy will receive a certificate of participation with special endorsements in areas where they contributed to their individual or team effort (such as schedules, mission design, CAD, or PDR).

The spring of 2019 saw the launch of L'SPACE Academy 2 with 198 students (56% the size of the fall 2018 Academy 1). Most of the attrition again was found to be due to schedule conflicts, and those students were allowed to join the summer L'SPACE Academy 2. Student teams were given detailed feedback about their L'SPACE Academy 1 PDR by the L'SPACE leadership team, and for L'SPACE Academy 2 the teams focused on polishing their PDR documents and running

Earth-based testing on their mission concept. Teams then made plans for Earth testing their mission concept via FaceTime or Skype. A final report was presented at the end of L'SPACE Academy 2.

We also ran another L'SPACE Academy 1 in the spring of 2019 to initiate a second cohort of students, and a vigorous recruiting campaign was launched through national science and engineering organizations such as the Society of Hispanic Professional Engineers (SHPE), Society of Women Engineers (SWE), and others. Two NASA education programs—the Minority Undergraduate Research and Education Program and the Space Grant Program—became partners with L'SPACE in disseminating the opportunity to their member institutions. The program received 633 applications (an increase of 64% from the fall Academy 1) and represented 253 institutions (an increase of 148%) from 40 states and Puerto Rico. The final enrollment had 38% minority students and 59% male, 38% female, and 2% other. In terms of disciplines represented, 60% were engineering students, 24% were science students, 10% were computer science students, 3% were mathematics students, and 3% were others. The spring 2019 L'SPACE Academy 1 reached its capacity (485 students) before the application deadline, and 148 students who applied after the Academy was full were encouraged to enroll in the summer 2019 Academy.

Our initial impression of student interest and participation in the L'SPACE Academy and the Space Works Project is tied to two main factors: (1) NASA is the sponsor, and (2) students are involved in authentic space-based STEM projects that are relevant and of interest to them (see Fig. 5.5). We are continuing to collect data on these programs through external evaluators to more deeply understand if students are gaining relevant workforce development skills that will aid them in preparing for STEM careers.

**FIG. 5.5**

Students provide real-time input during the L'SPACE sessions. *Dann Garcia.*

## Tips for Future Space-Based Authentic STEM Learning Initiatives

Based on our experience developing and implementing MSIP, Space Works, and the L'SPACE Academy, we offer the following recommendations to STEM education professionals and others seeking to nurture the future space and STEM workforce:

- **Students must be able to access and use real scientific data and professional, industry-standard tools in their research.** Using the same data as scientists, rather than relying on prepackaged data in worksheets, enables students to analyze and understand problems in context. MSIP data come from NASA Mars missions and are provided in the JMARS tool developed to aid scientists in their research. The JMARS tool is particularly valuable as well because one can download it for free, which enables its accessibility by students from all socioeconomic situations. Another aspect of accessibility is knowledge of how to use the tools, and MSIP provides teachers and students with training in use of JMARS to help them succeed in the implementation or completion of their projects. The MSIP team and educators have reported all types of students, from highly performing students to those who struggle with STEM content, going beyond their usual classroom knowledge and behavior and having learned to use JMARS to participate actively in authentic research.

  JMARS is also an essential tool for the Space Works and L'SPACE programs. The higher education teams access the JMARS data for planning their science experiments and surface operations for their mission concept spacecraft on a variety of bodies across the solar system. The knowledge and use of this NASA tool also gives their resumes a boost.

- **Mission scientists and engineers must be actively engaged in the authentic STEM experiences.** Many NASA mission team members are excited about education; however, the demand on their professional time limits how much they can support education. With forward planning, authentic STEM instructional designers can work with NASA education and outreach liaisons to look at ways to integrate subject matter experts and their knowledge through a variety of interactions and contributive pathways that will bring the relevancy of exploration to students. Contributions may be laid out as a "menu of options" to allow scientists and engineers to opt in where they have time and expertise. These options may include helping with the design of the experience, review of the science/engineering included in the experience to validate that it is correct and appropriate, synchronous or asynchronous participation within the experience, participation as a guest speaker during educator trainings, or helping to prepare answers to questions frequently asked by students.

  By leveraging technology platforms for MSIP, we reduced the time and effort for scientists and engineers to physically travel to locations to interact with students. We found that it removed a barrier for their involvement, and we had a better chance of drawing them into participating within the authentic STEM experience. For MSIP, the Mars science community continually contributed to curating the image files of archived data that were accessible to the student teams, supported the design and redesign of JMARS, and listened and guided the student research teams with feedback in their final MSIP reports. Subject matter experts (SMEs) were a major factor in Space Works and L'SPACE. Given the technical nature of their mission concept projects, SMEs provided "just in time" learning and a resource for technical Q&A in the weekly online or classroom settings. This was especially important for students from underserved communities, as in most cases, this type of technical mentoring was not available.

- **Students need basic information before they can begin to ask meaningful research questions and develop testable hypotheses that will guide their research.** For MSIP, the *Mars Image Analysis* activity introduces students to the types of images they will use, acquaints them with basic surface features on Mars (craters, volcanoes, gullies, canyons, lava tubes, etc.), teaches relative aging techniques, and instructs how to take quantitative data from images. This is critical background information and allows students to begin to formulate questions (see Fig. 5.6). In the Space Works and L'SPACE programs, this information was delivered via the weekly SME speakers. It is very important in the higher education programs that the SMEs were recruited to speak in a choreographed manner to deliver the information to coincide with the development of their mission concept.

- **Students must plan research investigations based on research questions and testable hypotheses they develop.** This is a critical part of authentic science experiences. When students are doing this, they are engaged in all of the STEM education standard practices and workforce development skills. Performing background research using published literature allows them to familiarize themselves with what is currently known. From there they can construct their own research, deciding what questions to pursue, what data to collect, what tools to use, and how they will use the data to answer their research questions and support or refute their hypotheses. This process was especially enlightening for the cross-disciplinary STEM teams in all our programs. Understanding the science drivers of their mission concepts informed the engineering design, and the optimizing of the project to the fullest extent was possible.

- **Teachers must be provided support throughout the process.** Few teachers have done scientific research and know the general process, let alone the tools and process of particular research areas. MSIP provides teacher guides that identify connections to key standards and the instructional objectives of the lesson. MSIP also provides documentation that shows the alignment of the lesson instructional objectives and learning outcomes. Teachers also receive one-on-one support from the MSIP staff in the form of emails, phone calls, and online video-conferencing as needed. If funding is available to support staff, it would be very useful to provide teachers with an in-depth online training on all aspects of the proposed program.

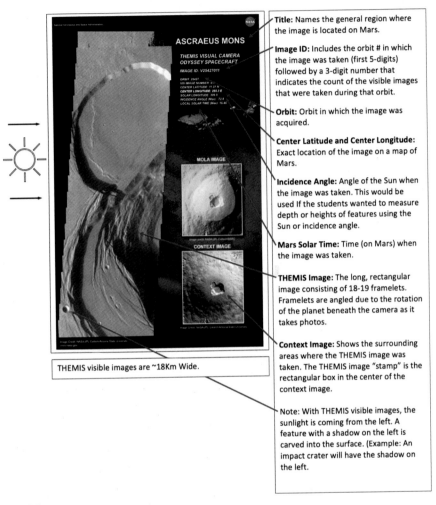

**Title:** Names the general region where the image is located on Mars.

**Image ID:** Includes the orbit # in which the image was taken (first 5-digits) followed by a 3-digit number that indicates the count of the visible images that were taken during that orbit.

**Orbit:** Orbit in which the image was acquired.

**Center Latitude and Center Longitude:** Exact location of the image on a map of Mars.

**Incidence Angle:** Angle of the Sun when the image was taken. This would be used If the students wanted to measure depth or heights of features using the Sun or incidence angle.

**Mars Solar Time:** Time (on Mars) when the image was taken.

**THEMIS Image:** The long, rectangular image consisting of 18-19 framelets. Framelets are angled due to the rotation of the planet beneath the camera as it takes photos.

**Context Image:** Shows the surrounding areas where the THEMIS image was taken. The THEMIS image "stamp" is the rectangular box in the center of the context image.

**Note:** With THEMIS visible images, the sunlight is coming from the left. A feature with a shadow on the left is carved into the surface. (Example: An impact crater will have the shadow on the left.

THEMIS visible images are ~18Km Wide.

**FIG. 5.6**

Example of MSIP background scaffolding for students and teachers. *ASU Mars Education Program.*

## Tying It All Together

The most effective ways to learn must evolve with the times. Learners too are evolving in the way they acquire information. The opportunities we, as space science educators, design must stay in step with our audiences' needs and preferences. From our experience over the last two decades in developing space-based, authentic STEM programming, we find the wonder and unknown nature of space can be used as an effective vehicle for engagement, but more than that, it can catalyze and invigorate learners to move toward higher, more complex understanding of our world and the value of STEM in society. Learners today do not just want to *know* about what is happening in space, they want to participate! Authentic STEM is a powerful bridge to bring learners into "doing" space. Combining participatory STEM opportunities, the best practices in education, and workforce development skills will enable learners to emerge from the STEM pipeline at the end of their terminal degree as practiced problem solvers ready to tackle the challenges of the future, whether in space or other science and technology disciplines.

## References

American Association for Community Colleges - https://www.aacc.nche.edu/research-trends/fast-facts/.

Burnham, R. (2010). *Middle school project discovers cave on Mars*. ASU News Now. https://asunews.asu.edu/20100617_skylight.

Committee on STEM Education National Science and Technology Council. (2013). *Federal science, technology, engineering, and mathematics (STEM) education 5-year strategic plan.* http://www.whitehouse.gov/sites/default/files/microsites/ostp/stem_stratplan_2013.pdf.

Klug Boonstra, S., & Christensen, P. R. (2013). Mars student imaging project: Real research by secondary students. *Science, 339*(6122), 920−921. https://doi.org/10.1126/science.1229849. http://science.sciencemag.org/content/sci/339/6122/920.full.pdf?sid=df98b8de-5cb7-47f0-8b9f-d195d1768618

National Academy of Sciences, National Academy of Engineering, and Institute of Medicine. (2007). *Rising above the gathering storm: Energizing and employing America for a brighter economic future.* Washington, DC: The National Academies Press. https://doi.org/10.17226/11463

National Research Council. (2005). *How students learn: History, mathematics, and science in the classroom.* Washington, DC: National Academies Press.

National Research Council. (2007). *Taking science to school.* Washington, DC: National Academies Press.

National Research Council. (2012). *A framework for K-12 science education: Practices, crosscutting concepts, and core ideas.* Washington, DC: The National Academies Press. NGSS Lead States. 2013.

National Research Council (NRC). (2012). *A framework for K−12 science education: Practices, crosscutting concepts, and core ideas.* Washington, DC: National Academies Press.

NGSS Lead States. (2013). *Next generation science standards: For states, by states.* Washington, DC: The National Academies Press. https://doi.org/10.17226/13165

Partnership for 21st Century Learning, A network of battelle for kids. http://www.battelleforkids.org/networks/p21/frameworks-resources.

President's Council of Advisors on Science and Technology. (2010). *Prepare and inspire: K-12 education in science, technology, engineering, and mathematics (STEM) for America's future.* https://obamawhitehouse.archives.gov/sites/default/files/microsites/ostp/pcast-stem-ed-final.pdf.

Tanenbaum, C. (2016). *STEM 2026: A vision for innovation in STEM education.* Washington: Office of Innovation and Improvement, US Department of Education. http://www.air.org/system/files/downloads/report/STEM-2026-Vision-for-Innovation-September-2016. pdf. (Accessed 22 January 2019).

The Mars Student Imaging Project (MSIP): https://marsed.asu.edu/msip.

WestPoint. (May 2010). *US Military Academy commencement address by Barack Obama.* NY: WestPoint.

# The Symbiosis of the Space Sciences and Popularization Activities

**Ron Miller**

A compelling argument can be made for space science being unique among all the sciences in that its roots lay deeply embedded in the arts.[1] At the same time, the space sciences have inspired works of art and cultural artifacts and activities for centuries. This symbiotic relationship between space science and the arts—and other activities aimed at communicating and popularizing space, thereby making it understandable and accessible to all people—is something that has remained in place to this day. Indeed, space exploration had its defenders and popularizers long before it became a reality. And, as I suggest, an argument can be made that space flight became a reality *because* of those who advocated and proselytized it and incorporated it into the arts and popular culture.

Space exploration has its active popularizers and place in the arts and popular culture today, all part of an unbroken continuum from earlier times. The challenge and irony today, however, is that as space exploration has become commonplace, the urgency remains for space popularizers to remind people of its importance. Starting with a brief look at the history of popularization that enabled space exploration and today's space popularizers to reach their current stature, this chapter examines the landscape of and continued need for popularization for space science pursuits to remain relevant to the American public and global publics also in the 21st century.

## Early Space Popularization Achievements

From the moment in 1610 when Galileo Galilei first turned a telescope toward the night sky, humans knew that Earth was neither unique in the universe nor alone. With Galileo's discoveries coming as they did during a time when new "worlds" were being discovered and explored on Earth, it did not seem to be a great leap of imagination to suppose that the new worlds discovered in the heavens would

---

[1] Space science is defined in this chapter to encompass all forms of space exploration, whether performed by humans in person or at a distance via human-controlled instruments. It is used somewhat interchangeably with "space exploration," "space flight," and "space travel" as much of the earliest instances of space popularization grew out of their progenitors' aspirations for people to reach space themselves.

**Space Science and Public Engagement.** https://doi.org/10.1016/B978-0-12-817390-9.00013-0

also someday see their explorers and colonists. Within a relatively short time, the world saw the publication of John Wilkins' *A Discourse Concerning a New World and Another Planet* (1638), Francis Godwin's *The Man in the Moone* (1638), Cyrano de Bergerac's *A Comical History of the States and Empires of the Worlds of the Moon and Sun* (1687), and many others in the following century. Likewise, the invention of the balloon in 1783 enabled humans to free their bonds to the Earth's surface. If a device could carry passengers through the air a few miles across the countryside of France, why not also to the Moon? Indeed, some of the earliest space travel tales of the first half of the 19th century depended on balloons to achieve their celestial goals. Rudolf Erich Raspe's *Gulliver Revived* (1786) was published only 3 years after the Montgolfier brothers' first flight and included a trip to the Moon by balloon. Edgar Allan Poe described a journey to the Moon via balloon in his story, "The Unparalleled Adventures of One Hans Pfaall" (1835), which anticipated in many details the stratospheric balloons of the 1930s and 1950s.

Shortly thereafter, Jules Verne's classic novel, *From the Earth to the Moon* (1865), and its sequel, *Round the Moon* (1870), showed for the first time that space travel was a matter of not only engineering and technology but also *known* engineering and technology. The novels did not suppose any imaginary propellants, materials, or devices such as antigravity. Verne was also the first author to convey the problem of interplanetary travel on a mathematical basis.

It is hard to overestimate the influence of Verne's novels. Not only did they directly inspire a virtual cottage industry of interplanetary storytellers, but also they influenced some of the seminal figures who helped create modern astronautics. Hermann Oberth, Wernher von Braun, and Robert Goddard, all of whom claimed to have been inspired by reading Verne's books, would go on to become space flight pioneers and would team with still others to popularize their aspirations in space.

One avid rocket enthusiast, Willy Ley, recognized that public support was an absolute necessity. Inspired by reading Hermann Oberth's book, *Die Rakete zu den Planetenräumen* (*The Rocket into Interplanetary Space*, 1923), Ley wrote a popularization of it, *Die Fahrt ins Weltall* (*The Journey into Space*, 1926). The following year he cofounded the Verein für Raumschiffahrt (the Society for Space Travel, or VfR). Among the VfR's earliest and most enthusiastic members was a teenaged Wernher von Braun. Over the next several years, Ley continued to write books about space flight while at the same time editing the VfR's journal, *Die Rakete*. He was also an active participant in the development of the VfR's experimental liquid fuel rockets. But after the German army took over the VfR and its work prior to the outbreak of the war, Ley fled to the United States. There he continued to proselytize rocketry and space travel in a series of magazine articles and popular books that included his 1949 collaboration with artist Chesley Bonestell, *The Conquest of Space*.

The book featured a text by Ley and 48 paintings by Bonestell, who had become the visual spokesperson for space travel after publishing an artistic tour among Saturn's satellites in *Life* magazine in 1944. The realism of Bonestell's artwork in conjunction with Ley's confident expertise (and astronomical facts double-checked by astronomer Dr. Robert Richardson) led the book to become a bestseller, going through nine printings within the following decade and convincing an entire generation of post-World War II readers that space flight was possible in their life-time. There are countless professional aerospace engineers and scientists working today who decided their careers when they read *The Conquest of Space* as children. Among its fans was a 32-year-old Arthur C. Clarke, who was then working on his first book. "To many," he wrote, "[*The Conquest of Space*] will for the first time make the other planets real places, and not mere abstractions. In the years to come it is probably destined to fire many imaginations, and thereby to change many lives" (Newell, 2019).

Three years later, *Collier*'s magazine sponsored a gathering of the world's great-est space experts who, in a series of illustrated articles, outlined one of the first comprehensive scenarios for the exploration of space. These experts included von Braun, Ley, physicist Dr. Joseph Kaplan, space medicine authority Dr. Heinz Haber, international law expert Oscar Schachter, astronomer Dr. Fred Whipple and, as illus-trators, Bonestell, Fred Freeman, and Rolf Klep.

The first issue of the *Collier's* space flight series appeared on March 22, 1952. Over the next 2 years, five more installments appeared in what has come to be known as "the *Collier's* space program." Virtually every aspect of space flight was covered: astronaut training, unmanned satellites, space stations, lunar landings, and a mission to Mars. Even the international legal questions involved in space flight were considered. According to von Braun and his fellow experts, the United States could have an artificial satellite in orbit by 1963, a 50-man expedition to the Moon by 1964, and a manned mission to Mars soon afterward. The first two parts of the program were expected to cost a "mere" $4 billion. Von Braun was entirely serious. The technology existed to carry out his plans, he insisted, however grandiose seemed. As von Braun explained,

*While the [Collier's] designs may be a far cry from what Mars ships some thirty or forty years from now will actually look like, this approach will serve a worthwhile purpose. If we can show how a Mars ship could conceivably be built on the basis of what we know now, we can safely deduce that actual designs of the future can only be superior. Only by stubborn adherence to the engineering solutions based exclusively on scientific knowledge available today, and by strict avoidance of any speculations concerning future discoveries, can we bring proof that this fabu-lous venture is fundamentally feasible.*

**(von Braun, 1956)**

These magazines hit the American public of 60-odd years ago like a bombshell. Until their appearance, the whole concept of space flight had been something rele-gated to the far future, if not strictly the stuff of science fiction. Suggesting that space flight was actually quite possible, the *Collier*'s series marked the beginning of the golden age of space flight—that period during which the American public showed a fascination, enthusiasm, and support for space flight it had never shown before—the astronautical equivalent of the aviation craze of the 1920s and 1930s. The von

Braun—Bonestell spacecraft appeared in thousands of incarnations in toys, advertisements, movies, television—everything seemed to have a rocket or space theme, no matter how unlikely the connection. That the taxpayers themselves had become so space-happy thanks to the *Collier's* articles went a long way toward enabling von Braun to convince US government officials to invest in an American space program.

During the heyday of the space race, there was a bevy of authors who specialized in popularizing rocketry and space flight. Frederick I. Ordway's *Annotated Bibliography of Space Science and Technology* listed 18 titles related to astronautics published during the 1940s (Ordway, 1962). By 1961, Ordway's list had swollen to 58 new books, not including the plethora of books destined for popular consumption, especially for younger readers. Working scientists such as Isaac Asimov, Hal Clement, Arthur C. Clarke, and Robert Richardson not only created fiction that proselytized space flight but also wrote nonfiction articles and books that explained space science to readers across a wide span of ages. Ley and von Braun also used science fiction as a means of communicating ideas and excitement about space travel. But scientists writing for science fiction readers are largely preaching to the choir. No one really needs to sell science fiction fans on space travel. Authors such as Martin Caidin and Lloyd Mallan pumped out entire shelves of books on the subjects, all of them for a lay audience and many aimed squarely at younger readers.

The dawning of the Space Age emerged in still other forms of popular culture. Though hardly the first film to be based on the theme of space flight, *Destination Moon* (1950) was a landmark in space travel movies in its attention to detail and accuracy, garnering almost universally positive reviews and an Academy Award for its special effects.[2] A big budget science fiction movie produced by George Pal, *Destination Moon*, was the first such film in the United States to examine human flight to the Moon in a critical light, exposing the potential dangers and challenges of sending astronauts to the Moon and bringing them back safely.

Not only the big screen but also the smaller, flickering screen at home was occupied by space themes. Among some of the earliest television programs developed for the younger members of the audience were science fiction shows set against a background of space travel such as *Space Patrol* (1950—55), *Tom Corbett—Space Cadet* (1950—55), and *Rocky Jones—Space Ranger* (1954—55). As happened with magazine articles and movies, it was not long, however, before reality caught up with these early television-based space adventures. When *Men Into Space* (1959—60) appeared on prime-time television, the difference between it and *Space Patrol* and *Tom Corbett* could not have been more striking. Done in a starkly documentary style, the program had the cooperation of the US Air Force as well as the talents of Bonestell, who designed the series' spacecraft, which in turn were heavily influenced by

---

[2] Other notable films of the first part of the 20th century included *Le Voyage dans la Lune* (*A Trip to the Moon*, 1902) by stage magician-turned-film maker Georges Méliès and *Frau im Mond* (*Woman in the Moon*, 1929), directed by Fritz Lang with technical advice from Hermann Oberth and presenting in authentic detail the construction and flight of a rocket to the Moon.

the work of von Braun. Even the television series *Star Trek*, when it premiered in 1966, depicted space flight realistically and derived its premise and stories from the principles of space science.

Indeed, by the time astronauts were journeying into space in the 1960s, the American public was regularly exposed to space exploration, both in fact as reported in the news media as well as in popular culture, albeit one that no longer depicted fictional ray-gun-wielding space explorers seen fighting monsters from the planet Mercury but instead reflected scientists and engineers contending with more realistic problems, such as attempts to refuel an orbiting spacecraft or the construction of a space telescope. Reality and popular culture helped propel the continuation of the other throughout the decade of the United States–Soviet space race and together spurred the excitement many Americans and people around the world felt when astronauts reached the Moon in 1969.

## Space Popularizers: The Importance of Inspirational Figures Today

Fast forward to today, and the problem of maintaining public enthusiasm for space exploration can be seen as a repetition of the history of aviation. In 1927, Charles Lindberg was lionized following his solo flight across the Atlantic. His return to the United States was celebrated with a massive ticker tape parade in New York City. The country was aviation crazy. But within a few decades of Lindberg's epochal achievement, thousands of people were routinely crossing the world's oceans by air without thinking twice about it and without making headlines. What had been extraordinary for one person became the norm for multitudes.

Much the same thing has happened to space exploration. Between the end of World War II and the first human space flights, there was frenzy associated with space travel that equaled that of prewar aviation. References to space flight were ubiquitous, ranging from serious books and articles in national magazines to toys, games, and even wallpaper.

The irony lies in the fact that it is probably not possible to maintain a high level of enthusiasm for something that has become commonplace. While no one would argue that space travel has become as routine as transatlantic aviation, we are in the midst of the Space Age and no longer at its threshold. There are few great "firsts" to claim newspaper headlines. Making space flight an ordinary, everyday occurrence had once been the dream of people like Ley, von Braun, and Clarke, but few of them thought about the price that would have to be paid. With ordinariness comes complacency.

While we may never again see anything like the frenzied enthusiasm for space exploration that occurred in the 1950s and 1960s, this is not to say that the public has become completely jaded. People cheered at the inaugural Space Shuttle launch in 1981 as well as at the vehicles' retirement in 2012. There was a hint of an ember of enthusiasm for space exploration still being alive in the public breast when the *Path-finder* Mars rover became one of the media heroes of 1997. For weeks, the official NASA *Pathfinder* website was receiving more hits than any other site on the planet. It was no small feat for *Pathfinder* to have inspired some of the first new space-related toys in nearly 40 years: even the Space Shuttle did not receive the honor of a Hot Wheels replica! The American public was galvanized by the solar eclipse visible across the country in 2017. Most recently, the flight of Doug Hurley and Bob Behnken aboard Space X's Dragon capsule to the International Space Station—the first of American astronauts aboard a privately owned and operated spacecraft—captured extraordinary traditional and social media interest.

These events received the publicity they did thanks to concerted efforts on the part of NASA, Space X, and others to get them media attention. Without doubt, public relations offices of space-related organizations have kept the firsts in front of people around the world, whether reaching out to news media outlets or spreading the word via social media channels. But one of the factors critical in sharing these activities has been the involvement of inspirational figures—space scientists, engineers, astronauts, and others adept at public communication. Such individuals have been tapped to provide commentary about space-related events as they have unfolded to give the public an in-depth look into the scientific and technical dimensions as well as to explain their historical and cultural significance. Some popularizers of space science and exploration have emerged who have sought to bring information and excitement to the public around specific space missions and space science more generally.Indeed, beyond Wernher von Braun's heyday, starting in the 1970s Carl Sagan, became a luminary and champion for space exploration and invited the public to join him through his award-winning television series *Cosmos* and through his numerous books in the wonder of discovery across our solar system and beyond. Within the sciences alone, several other respected individuals—Neil deGrasse Tyson, Alan Stern, Carolyn Porco, Steve Squyres, and Bill Nye, to call out a few—have made a special effort to communicate with the public and inspire people about the possibilities of space exploration. And they all have in common a means for communication with the public denied to earlier proselytizers: the Internet.

Perhaps the most visible is Tyson, who in attempting to don the mantle once worn by Carl Sagan has made a second career of becoming the public face for astronomy and space exploration. An astrophysicist, Tyson is the Frederick P. Rose Director of the Hayden Planetarium at the Rose Center for Earth and Space, which is part of the American Museum of Natural History in New York City. In 2014, he started as host to a sequel to Sagan's *Cosmos*. The following year, the National Academy of Sciences awarded him its Public Welfare Medal for his "extraordinary role in exciting the public about the wonders of science" (National Academies, 2015). He is a regular presence on television and radio, appearing on venues ranging from the "Colbert Report" to the History Channel to National Public Radio. Focusing his career as a scientist and as an advocate for science education in the field of astronomy, he has been an outspoken supporter of NASA, arguing that "the most powerful agency on the dreams of a nation is currently underfunded to do what it needs to be doing." In a 2012 appearance before the US Senate Science Committee, he observed that "… NASA's annual budget is half a penny on your tax dollar. For twice that—a penny on a dollar—we can transform the country from a sullen, dispirited nation, weary of economic struggle, to one where it has reclaimed its 20th century birthright to dream of tomorrow" (Tyson, 2012).

Best known today as the principal investigator of the New Horizons mission to Pluto, Alan Stern is an engineer and planetary scientist by training. Listed among *Time* magazine's "100 Most Influential People in The World" in 2007, Stern has been indefatigable in his efforts to inspire a fascination for the exploration of the solar system's outer limits. He lectures widely and is the coauthor of books such as *Chasing New Horizons: Inside the Epic First Mission to Pluto* (Stern & Grinspoon, 2018) and *Pluto and Charon: Ice Worlds on the Ragged Edge of the Solar System* (Stern & Mitton, 1997).

Like that of Alan Stern, Carolyn Porco's name is inextricably bound to a spectacular example of interplanetary exploration. As the leader of the imaging science team on the Cassini mission which orbited Saturn from 2004 to 2017, she not only has been instrumental in publishing fundamental discoveries about Saturn and especially its enigmatic satellite, Enceladus, but also has been active in sharing her enthusiasm for space exploration with the public at large. Articles under her name have appeared in popular publications such as the *London Sunday Times, New York Times, Wall Street Journal, Guardian, Arizona Daily Star, Sky and Telescope, Scientific American, American Scientist*, and *Astronomy* magazine. Not only has she prepared articles for the PBS and BBC websites, but she has also maintained a website especially devoted to the images and discoveries made by the Cassini orbiter: ciclops.org. Most importantly, this site is designed to be easily consumed and understood by the layperson, whom she addresses directly via her "Captain's Log" page. Of perhaps special interest is the fact that Porco has devoted an entire section of the Ciclops website to artwork depicting Saturn and its moons. Space artists—professional and amateur—from all over the world have contributed, and continue to contribute, dozens of dazzling images to this collection. She has even sponsored a competition for composers to create works of music inspired by the Cassini mission. Porco was selected in 1999 by the *London Sunday Times* as one of 18 scientific leaders of the 21st century and by *Industrial Week* as one of "50 Stars to Watch." In 2009, *New Statesman* named her as one of the "50 People Who Matter Today." In 2010, the American Astronomical Society awarded her the Carl Sagan Medal for Excellence in Public Communication in Planetary Science. And in 2012, Porco was named by *Time* magazine as one of the 25 most influential people in space.

Steve Squyres, a principal investigator of NASA's Mars Exploration Rover mission, is another planetary scientist who has applied energy in promoting space exploration. ABC News featured Squyres as its Person of the Week for January 9, 2004, just a few days after the *Spirit* rover's successful landing on Mars, and "World News Tonight" anchor Peter Jennings declared that Squyres "has gotten us all excited" about the exploration of Mars. Squyres received the 2005 Rave Award for science by *Wired* magazine for overseeing the creation of the *Spirit* and *Opportunity* rovers. Squyres' book, *Roving Mars: Spirit, Opportunity, and the Exploration of the Red Planet* (2005), was the basis for the Disney IMAX documentary film, *Roving Mars*. Squyres was the recipient of the 2004 Carl Sagan Memorial Award and the 2009 Carl Sagan Medal for Excellence in Public Communication in Planetary Science.

One additional name should not be overlooked and that is Bill Nye, more familiarly known as Bill Nye the Science Guy, after the eponymous television series that ran from 1993 to 1998. While the program, which was an updated version of *Watch Mr. Wizard*, did not focus on the space sciences, Nye has subsequently taken on the position of executive director of The Planetary Society, a nonprofit organization devoted to public outreach and advocacy for topics related to the space sciences. He frequently hosts televised and online broadcasts surrounding notable space events.

Few science advocates have had the charisma and personality to enjoy the sort of celebrity usually accorded movie stars, but some have made their mark in popular media venues. For example, Tyson, along with NASA astronaut Mike Massimino, has made guest appearances on television sitcoms such as *Big Bang Theory* (2007—19) and, along with Apollo 11 astronaut Buzz Aldrin, has seen his animated counterpart on *The Simpsons* (1989—). Tyson and Nye have both appeared on multiple occasions on late night television shows. Earnestly and tirelessly, these individuals have worked to support and encourage public science education. While it is difficult to know the extent to which they are making a difference in attracting people who were not already supporters of space exploration, appearances like these at least continue to keep space science in front of a broad public audience.

## Space Science and the Arts

Lectures, televised science series, and nonfiction books all have their place in the popularization of the space sciences, but it is through the arts that popular culture finds its expression. This often takes the form of entertainment, which might include film, music, television, video games, comic books, toys, advertising, and even consumer products, sports, and fashion.

The visual arts have had a long tradition in both being inspired by and inspiring the space sciences. Ever since the first Chesley Bonestell paintings were published in the 1940s, art has been a major factor in communicating the excitement of space exploration to the general public. As we are a visually oriented species, the arts have an emotional impact difficult to achieve by any other media. We have already discussed the profound effect Bonestell's artwork for the *Collier's* magazine series on the future of space exploration had on the both public perception of space travel and on the development of space exploration itself.

Space artists themselves have often taken matters into their own hands. Instead of providing illustrations for books written by others, many have become author-illustrators, creating both the illustrations and the text for books about space, space travel, and astronomy. Michael Carroll, one of the founding members of the International Association of Astronomical Artists (IAAA), is the award-winning author of more than a dozen books, including *Alien Seas* (2013) and *Alien Volcanoes* (2008). He has written two science fiction novels based on the future exploration of the solar system as well as *Space Art: How to Draw and Paint Planets, Moons and Landscapes of Alien Worlds* (2007). Artist Mark Garlick has written and illustrated 15 books about astronomy including *Cosmic Menagerie* (2014) and *The Story of the Solar System* (2018). David Aguilar has written and illustrated 11 books about astronomy and space travel, including several for younger readers, such as *Little Kids First Big Book of Space* (2012).

Other artists have taken a different tack from that of the visual artists in presenting the excitement of exploring other worlds. While there were once hit recordings ranging from "Telstar" to "Rocket Man," the list of music inspired by astronomy and the space sciences has expanded over the years, as a compilation by Andrew Fraknoi reveals (Fraknoi, 2018). In one example from recent times, Australian-born composer Amanda Lee Falkenberg has composed a full-length symphony, *The Moons*. Each of its seven movements is devoted to a different major natural satellite, from Saturn's Titan to Earth's Moon. In the process of creating the symphony, she consulted numerous planetary scientists, such as Dr. Ashley Davies from NASA and Dr. Ralph Lorenze from Johns Hopkins University's Applied Physics Laboratory. Falkenberg describes the work as celebrating "the merging of art and science to inspire and educate wider communities at large and to provide a theatrical foundation for the thrilling stories of these moons and the spectacular discoveries being made in our outer solar system" (Falkenberg, 2020).

Hollywood, meanwhile, has begun to take a distinct about-face regarding the depiction of space exploration. Where space was once only a convenient background for alien invasions and interplanetary battles, more and more films are being produced that depict space exploration realistically and with more than a passing nod to good science. Films such as *Contact* (1997), *Moon* (2009), *Gravity* (2013), *Europa Report* (2013), *Interstellar* (2014), *The Martian* (2015), and *First Man* (2018) all based their stories on situations that evolved from a realistic scientific premise or from the environment of another world. In other words, science is an active participant in these films instead of merely a background extra.

It is not unusual to see a reputable scientist or engineer taking an active part in the production of a Hollywood space film, something that had happened in the past only with *Die Frau im Mond* (1929), *Destination Moon* (1950), and *2001: A Space Odyssey* (1968). *Contact* was notable for not only having been based on a novel by a distinguished scientist and astronomer, Carl Sagan, but Sagan and his wife, Ann Druyan, were credited as coproducers. More recently, astrophysicist Kip Thorne advised the producers of *Interstellar*, which involved the realistic depiction of a black hole and exoplanets as well.

*The Martian* is a particularly good example of the trend toward realism. It had already been a *New York Times* best-selling novel, meaning that it had been widely read outside the narrow confines of science fiction fandom. Turned into a motion picture, it has earned more than a half-billion dollars at the box office. What makes this movie especially significant is that the problems facing the film's hero are evolved directly from the environment of Mars itself and are eventually resolved by the application of science and engineering. Movies like this can give viewers a sense of what it would take to survive on the red planet while keeping these distinct future possibilities in the public mind.

## Connecting to Space Science Online

The Internet straddles the gap between pure popular culture and the deliberately educational. One can find, on the one hand, serious TED Talks and NASA programming, and on the other, vintage movies and cartoons. The Internet is the obvious choice in the current era for communicating ideas about space exploration and to engage the public's interest. And the sheer volume of websites devoted to the subject bears this out. A search I conducted for "space science" estimates more than 3 billion sites found on the entire Internet. A search for "space science" videos yields 227 million titles, including those on YouTube. (Of course, many duplicate entries push up the counts.)

There are some extraordinarily effective, well-designed, well-written, and well-informed websites devoted to promoting and informing the public about space science and space travel. Space.com, for instance, is one of the oldest and best respected of these. Founded in 1999, it covers the latest discoveries, missions, trends, and future ideas about the exploration of space. And NASA's website (www.nasa.gov) is the go-to site for anyone searching for information about current missions and projects as well as the history of space flight and future missions. The site is deep, covering every level of interest, from the established space enthusiast to the grade-school student. NASA's site https://solarsystem.nasa.gov, for instance, gives browsers an armchair tour of the planets and other bodies of the solar system.

In addition to the bevy of websites created by NASA, many excellent websites are from private sector companies. For example, Mars One (https://www.mars-one.com/), which has the ambitious goal of sending humans to Mars, states that its crews "will go to Mars not to simply visit, but to live, explore, and create a second home for humanity. The first men and women to go to Mars are going there to stay." Whether their plans are realistic or not, the science and technology are sound, and for that reason, Mars One might be thought of as the modern equivalent of von Braun's *Collier's* effort. Blue Origin (https://www.blueorigin.com/), a company founded by Amazon CEO Jeff Bezos, makes its goals clear from the outset: "Blue Origin believes in a future where millions of people are living and working in space … Blue Origin is committed to building a road to space so our children can build a future." The company's "New Glenn" reusable launch vehicle will be specifically designed to carry passengers and payloads routinely to earth orbit and beyond and, as the company boasts, "will build a road to space." Virgin Galactic's site (https://www.virgingalactic.com) targets the general public specifically and is devoted to inspiring an interest in future space travel. As the site points out, "There is little that compares to the sense of awe that takes hold as we raise our eyes to the night sky. Despite the fact that millions of people would love to experience space, fewer than 600 have been given that opportunity." To this end, Virgin Galactic has taken a proactive role in inspiring and educating young people. Through scholarships and grants, Virgin Galactic's Galactic Unite (https://www.galacticunite.com/) and its Future Astronauts program hopes to exploit "the power of space to support and inspire young people to pursue an education in STEM (Science, Technology, Engineering and Mathematics)."

Unfortunately, the best websites are found alongside a multitude of sites that are uninformed, out of date, inaccurate or, worst of all, based on not just bad science but pseudoscience. For instance, a Google search for "moon hoax" in 2019 resulted in nearly 3 million videos. Many pseudoscientific YouTube channels are very professionally done, sometimes even more impressively so than videos produced by NASA, universities, or other legitimate organizations, seemingly bearing the hallmarks of a trustworthy source of information to the unwary. A search engine such as Google provides at least a modicum of filtering: Google will display the most popular sites, which are often, by default, the most reputable ones. For instance, the top 10 results from a search for "travel to Mars" are all NASA sites or sites of legitimate news sources. YouTube has no filters to sift the wheat from the chaff: anyone capable of creating and uploading a video is welcome to do so. The popularity of a website or video is obviously no guarantee of its veracity or usefulness: a YouTube video that claims that gravity itself is a hoax has been viewed more than 8,000 times (https://www.youtube.com/watch?v=Z9V5IiHN1fk).

With a generation that is more and more visually oriented and attuned to life online, there is great potential to reach young people—along with older populations—through engaging websites and videos. At the same time, the lack of filtering can be a real impediment to disseminating factual information and inspiring a concerted public interest in the space sciences.

## Space Science Experiences

Books, movies, websites, and social media are all effective means of conveying the excitement of the space sciences to the general public. But there are ways of achieving this excitement that may be more immediate, interactive, and involving means of popularizing space and bringing it to the people. NASA initiated its ambitious Young Astronauts program in 1984. It attracted 2 million children in 100,000 chapters in this country and abroad. It faltered, however, following the September 11, 2001, attacks in the United States, never fully recovered, and now exists as a program administered largely through elementary schools and online videos hosted by

NASA such as its "Televised Space Academy." Space Camp, at the US Space & Rocket Center in Huntsville, Alabama, has been successfully operating for more than 35 years, offering interactive experiences to young people, adults, and families. Out of the 750,000 who have attended, nearly a dozen "graduates" have gone on to become astronauts themselves. Space Camp remains a robust influence today on promoting continuing STEM education and student and public interest in astronautics.

People of all ages have also enjoyed space-related festivals in communities around the country. Every year for the past decade, the city of Tucson, Arizona, has played host to SpaceFest, a 3-day gathering of Apollo, Gemini, and Space Shuttle astronauts, space historians, guest speakers, authors, astronomers, films, and vendors supporting an array of STEAM (STEM plus the arts) events. The annual programming includes a full-scale art show organized by the IAAA featuring space art from artists from all around the globe. Many of the events are part of the paid admission, but many are free and open to the general public. According to its organizers, Novaspace, a Tucson space art gallery and memorabilia dealer: "Our goal is to reach out to those who like space, but just don't know it yet!" (Spacefest, 2019) Spacefest was originally developed by the late Kim Poor, Novaspace founder, a space artist himself and one of the founding members and first president of the IAAA.

Other communities, schools, museums, and other organizations have sponsored similar events, though not always on the same large scale. Reno, Nevada, was home to its own Space Fest in May 2018. The 4-day event was billed as the "Biggest Little Science + Fiction Film Festival in the World!" (Filmfreeway, 2019) Held at the Challenger Learning Center of Northern Nevada, it not only featured space-inspired films and documentaries but also was the venue for the annual Nevada Space Center Hall of Fame Awards, which honored women's contributions in space exploration. In September of that year, the Long Beach Comic Con and the Columbia Memorial Space Center in Downey, California, collaborated on Space Expo. The 2-day event was described as being "dedicated to providing high-quality STEM education made accessible through a pop culture atmosphere [bringing] together the space and entertainment industries in order to strengthen them by fostering interest and creating future scientists and engineers" (longbeach-comiccon, 2019). Among presentations at the Expo was "Women of Mars," in which several of the women who drive the NASA Mars rovers and analyze the information sent back to Earth talked about their experiences in STEM fields and the roles women play in modern space exploration.

Seattle's Museum of Flight hosted a Space Expo in 2018 that "combined the arts and rocket science to explore and interpret space exploration" (museumofflight, 2018). The day-long event included lectures, virtual reality experiences, and performance art with guests from NASA, the Mars Society, the University of Washington, and Media Art Xploration. The latest is a nonprofit organization based in San Francisco dedicated to "using performing and interactive arts to … build a synergy between the artist and the scientist" (mediaartexploration, 2019). In 2019, coinciding with the 50th anniversary of the Apollo 11 lunar landing, Media Art Xploration hosted MAX 2019: A Space Festival, which featured "live art, poetry slams, performance installations, cool scientists, and more!" (mediarrtexploration, 2019). Among these events was a futuristic fashion show, an original musical play about an elderly couple moving to another planet and an "Intergalactic Travel Bureau" that allowed visitors to plan their vacation on Mars.

These and similar events, both large and small, provide an immersive, hands-on experience for the general public that the Internet, print, film, or video finds hard to match. Most of these efforts exploited popular culture to a lesser or greater degree. Reno's Space Fest was built around a science fiction film festival, Seattle's Space Expo included performance art, and MAX 2019 included poetry and performances. While many such festivals are attended by those who already are familiar with space exploration and thrive on such experiences, studies show that science festivals can have significant impacts on learning and promoting future learning and engagement with science. These informal settings for science learning can be particularly beneficial to populations underrepresented in the sciences who would not otherwise have exposure to the sciences (https://www.informalscience.org/). These events also lend support to the artists, writers, musicians, and performers who have been inspired by space travel and the space sciences and can continue to play an important role in space popularization.

## The Future Need for Space Science Popularization

Popularization of space science and space exploration through a variety of cultural works and outspoken advocates started long before the Space Age began and was instrumental in helping to herald its arrival. As space flight became a reality, it remained a trope in everything from television shows to comic books to pencil sharpeners. By the time astronauts landed on the Moon in the late 1960s, a generation of scientists and engineers had taken inspiration from the symbiosis of the real-world achievements of NASA and the efforts of popularizers to incorporate space into the American imagination and many dimensions of life. It is difficult to extricate the relative roles fact and fiction had in producing the future.

Space is not the craze it was in popular culture of the 1960s. Today's world is cluttered with more television stations, types of media, and distractions than existed in that era. Space exploration also is no longer a novelty. Yet, it continues to permeate modern day movies, arts, and festivals and remains the passion of many scientists and communicators and everyday people who make it their business to explain and espouse it to lay audiences. NASA, its global counterparts, and the space industry continue to make advances that command the attention of news and social media, and public polls show that Americans generally remain favorable to having a robust national space program. A poll conducted by the Pew Research Center in 2018 showed that 72% of the 2500 people surveyed thought it was important that the United States continue to lead the world in space exploration, while 80% thought the International Space Station was a good investment (https://www.pewresearch.-org). The symbiotic relationship between space exploration and popularization activities has evidently continued into the modern day, even if we cannot always know what the public takes away from imbibing it.

Like any symbiotic relationship, that between the space sciences and popularization activities must continue to be a two-way street. The space community cannot let up its momentum and passively wait for movies, television, comic books, and toy manufacturers to pay attention—they need to be proactive and harness these resources. Even while human space reaps the lion's share of attention in popular culture, advocates have reason to be concerned. NASA has set big aspirations, undertaking the ambitious Artemis program, the goal of which is to land a man and a woman on the Moon by 2024. Meanwhile, the poll that showed majority support for a strong national space program revealed that only 18% of those polled thought that sending humans to Mars should be a top priority and just 13% strongly favored a return to the Moon (https://www.pewresearch.org).

Astronomers, planetary scientists, and other space scientists have even greater reason to remain proactive. Black holes, quasars, and other objects of astronomical interest have been far less often featured in popular culture and the news media than human space activities—clearly, as stand-alones they lack the romantic angle or plotline fodder that human space travel can supply. As a result, the space sciences are less apparent, familiar, and relatable to the general public. Notably, they receive only a fraction of the funding for research that human space flight receives.

Perhaps the most significant reason for continued popularization and cultural engagement with space science is the preservation of not only this scientific discipline as one of many. American students consistently lag the rest of the world in math and science. According to studies undertaken in 2015—16 by the Program for International Student Assessment under the aegis of the Organization for Economic Cooperation and Development, the United States ranked 31st among 70 nations in reading, science, and math (http://factsmaps.com). And while the number of graduate degrees earned in STEM fields has been trending upward, 55% of mathematics, computer science, and engineering students at American universities come from other countries (https://cgsnet.org). Meanwhile, an antiscience and antiintellectual trend exists in the United States, which has been in recent decades exacerbated by an increasing distrust of government. Accordingly, many regard scientists and their institutions as part of a grand, overarching conspiracy. Six percent of Americans (comprised more by millennials than any other generation) believe that the Apollo lunar landings were a hoax (https://www.ipsos.com). Beliefs like this are spread and encouraged by the ease and prevalence with which individuals can promote pseudoscience—from astrology and "chemtrails" to the flat Earth and antivaccination movements. YouTube alone has hundreds of thousands of videos, many of them of highly professional quality, dedicated specifically to undermining confidence in science and scientists. Many of the arguments put forward by these videos and websites can seem plausibly compelling to the uninformed layperson. Unchecked and unguided, popular culture can work against science as easily as it can work for it.

There is a lot of information and misinformation vying for public attention today. One potential tactic to counter the declining interest and faith in science is the encouragement of young people to take part in STEM programs. The other is to encourage the general public's enthusiasm and support for space exploration. The number of present-day astronomers, space scientists, and aerospace engineers whose origins were as *Star Trek* fans is legion. More attention and effort need to be focused on generating excitement, enthusiasm, and support for the space sciences among the general public and not just among its established fans. The job of inspiring the general public's support for space science and exploration rests with the scientists, engineers, and leaders in the privatization of space exploration—Elon Musk, Jeff Bezos, Richard Branson, and others—who, through progress and tapping into opportunities to capture the public imagination today, can energize public support for space exploration and build tomorrow.

## References

https://academickids.com/encyclopedia/index.php/Steve_Squyres (Accessed April 2019)

https://cgsnet.org/ckfinder/userfiles/files/Graduate%20Enrollment%20%20Degrees%20Fall%202015%20Final.pdf (Accessed March 2020)

http://content.time.com/time/specials/packages/article/0,28804,1860871_1860876_1860992,00.html (Accessed April 2019)

http://factsmaps.com/pisa-worldwide-ranking-average-score-of-math-science-reading/ (Accessed April 2019)

https://filmfreeway.com/SciOn (Accessed April 2019)

https://fivethirtyeight.com/features/americans-are-smart-about-science/ (Accessed April 2019)

https://iaaa.org/moon-io-6th-movement-of-the-moons-symphony-composed-by-amanda-leefalkenberg/ (Accessed March 2020)

https://longbeachcomiccon.com/space-expo/ (Accessed April 2019)

https://mediaartexploration.org/ (Accessed April 2019)

https://mediaartexploration.org/festival/max-2019-a-space-festival/ (Accessed April 2019)

http://static.c-spanvideo.org/files/pressCenter/C-SPAN-Ipsos-Poll-on-Attitudes-Toward-Space-Exploration.pdf (Accessed April 2019)

https://www.bls.gov/spotlight/2017/science-technology-engineering (Accessed April 2019)

https://www.edweek.org/ew/section/multimedia/chart-which-stem-jobs-will-be-in.html (Accessed March 2020)

http://www.fraknoi.com/wp-content/uploads/2018/06/Fraknoi-Music-and-Astronomy-Article.pdf (Accessed April 2019)

https://www.informalscience.org/sites/default/files/SFA_Yr_2_Summative_Evaluation.pdf (Accessed March 2020)

https://www.ipsos.com/sites/default/files/ct/news/documents/2019-07/c-span-space-exploration-07-10-2019_for_release.pdf (Accessed April 2019)

https://www.museumofflight.org/Plan-Your-Visit/Calendar-of-Events/4774/space-expo-2018 (Accessed April 2019)

https://www.pewresearch.org/science/2018/06/06/majority-of-americans-believe-it-is-essential-that-the-u-s-remain-a-global-leader-in-space/ (Accessed April 2019)

https://www.pewresearch.org/science/2019/03/28/what-americans-know-about-science/ (Accessed April 2019) https://www.spacefest.info/?page_id%bc5142 (Accessed March 2020)

https://www.statepress.com/article/2014/09/the-uncertain-future-of-space-exploration (Accessed April 2019)

https://www.tested.com/science/space/456983-art-and-proposed-50-man-lunar-expedition-1950s-colliers-magazine/ (Accessed April 2019)

https://www.themoonsmusic.com/ (Accessed March 2020)

https://www.tor.com/2012/04/13/neil-degrasse-tyson-on-the-nasa-budget/ (Accessed March 2020)

von Braun, W. (1956). *The exploration of Mars*. Viking Press.

National Academies. (2015). https://www.nationalacademies.org/news/2015/02/neil-degrasse-tyson-to-receive-public-welfare-medal-academys-most-prestigious-award.

Newell, K. (2019). *Destined for the stars*. University of Pittsburgh Press.

Ordway, F. I. (1962). *Annotated bibliography of space science and technology*. Washington, DC: Arfor Publications.

Stern, & Grinspoon. (2018). *Chasing new horizons: Inside the epic first mission to Pluto*. NY: Macmillan.

Stern, & Mitton. (1997). *Pluto and Charon: Ice worlds on the ragged edge of the solar system*. NY: Wiley.

Tyson, N. (2 March 2012). *Written testimony submitted to the U.S. Senate committee on commerce, science, and transportation. Russell senate office building, room 253.*

## Further Reading

A Cult of Ignorance. *Newsweek*, (1980).

Canto, C., & Faliu, O. (1993). *The History of the Future*. Paris: Flammarion.

Cooke, H. L. (1971). *Eyewitness to Space*. NY: Abrams.

Irvine, M. (2002). *Creating Space*. Collector's Guide Publishing.

Kitahara, T. (1989). *Yesterday's Toys, 3: Robots, Spaceships, and Monsters*. San Francisco: Chronicle Books.

Launius, R. D., & Ulrich, B. (1998). *NASA and the Exploration of Space*. NY: Stewart, Tabori and Chang.

Leiberman, R., & Ordway, F. I., III (1992). *Blueprint for Space*. Washington, DC: Smithsonian Books.

Mark Young, S., Duin, S., & Richardson, M. (2001). *Blast Off!* Milwaukie, OR: Dark Horse.

Miller, R. (2014). *Art of Space*. Minneapolis, MN: Zenith Press.

Miller, R. (2016). *Spaceships*. Washington, DC: Smithsonian Books.

Miller, R., & Durant, F. C., III (2000). *The Art of Chesley Bonestell*. London: Paper Tiger.

The Inspiration of Astronomical Phenomena. In Campion, N. (Ed.), *Culture and cosmos*, (2004). Trowbridge, UK.

Topham, S. (2003). *Where's My Space Age?* Prestel Publishing.

Tumbusch, T. N. (1990). *Space Adventure Collectibles*. Wallace-Homestead Book Co.

# Public Engagement With Planetary Science: Experiences With Astrobiology and Planetary Defense

**Linda Billings**

## Introduction

What are the goal, the aim, the objective, and the purpose of public engagement with planetary science? And what exactly is "public engagement?"

The American Association for the Advancement of Science's (AAAS) Center for Public Engagement with Science and Technology defines public engagement as "intentional, meaningful interactions that provide opportunities for mutual learning between scientists and members of the public." AAAS notes that public engagement "is closely related to and influenced by science communication, science education, and other fields," such as public outreach (Center, 2016).

I am a social scientist with a PhD in mass communication. As such, I cast a critical eye on efforts to promote public engagement with space science, assessing them in their broader cultural context. Since the 1990s, I have worked with a number of NASA science programs on communication strategy and planning, public education, and public engagement. I am currently working with NASA's astrobiology program and Planetary Defense Coordination Office. My primary justification for the planetary science communication and engagement efforts I have worked on is the need to fulfill NASA's statutory responsibility to "provide for the widest practicable and appropriate dissemination of information concerning its activities and the results thereof"[1] and ensure that decision-makers and citizens have access to clear, concise, correct, and timely information about the programs with which I work. I want the people who are paying for the research—taxpayers—to understand what science is, what it does, how it works, and what public interests and national objectives it serves.

The NASA program managers I have worked with by and large share these aims. It appears to me, however, that the primary aim of NASA's public engagement efforts at higher levels is to foster public interest, enthusiasm, and support for

---

[1] Section 203(a)(3), National Aeronautics and Space Act of 1958.

**Space Science and Public Engagement. https://doi.org/10.1016/B978-0-12-817390-9.00004-X**

NASA. NASA tends to measure the success of its engagement efforts by counting—headlines; teachers, students, and citizens participating in engagement projects; participants in tweet-ups and Facebook Live events; and so on. Such numerical assessments do not reveal much about the effects, or the benefits, of public engagement efforts.

While another of my goals in public engagement with planetary science is to determine what citizens care about and consider worthy of public funding, my experience has shown me that while NASA and the planetary science community writ large claim they are committed to public engagement, neither the agency nor the community is committed to determining how citizens envision the future of planetary exploration. The US National Academy of Sciences' decadal surveys of planetary exploration goals and objectives, which NASA uses to guide planning, draw on input from the science community but not from other citizens. The Academy's Space Studies Board held a 3-day workshop in 2010 called "Sharing the adventure with the public: the value and excitement of 'grand questions' in space science and exploration" (Space Studies Board, 2011).[2] A major focus of the workshop was on "ways to better sustain public understanding of, interest in, and involvement with NASA science and exploration efforts" (p. vii). I gave a talk at this workshop in which I said I was not convinced that more and better communication would increase public support for NASA. I also said that public participation in planning and policy making might be the only path to "enduring public involvement" (p. 4).

Most planetary scientists likely would say that planetary science has had a considerable public impact. However, research has not documented the nature, breadth, or depth of that impact. The same is the case for assessments of the breadth or depth of the impact of public engagement projects in planetary science.[3] The public impact of planetary science, or, alternatively, the public value of planetary science, is poorly understood, as little research has been published on the subject (Billings, 2019). Historian Steven J. Dick (2014) has noted that assessing the public impact of space science and exploration is a complex endeavor. "One can ask, for example, what does impact mean? Who is being impacted? What is the evidence that anyone is being impacted? And if there is an impact, individuals are undoubtedly affected in different ways depending on their worldviews or individual interests and predispositions" (p. 74). Assessing public impact, or public value, is a social scientific enterprise, and most space agencies, organizations, and companies do not have the proper expertise to conduct such research.

---

[2] I served on the planning committee for this workshop.

[3] A search of Google Scholar on October 16, 2018, for publications about the "public value of planetary science" or the "public value of planetary exploration" yielded no results. A search of Google Scholar for publications about the public value of science yielded 473 results. A search of the US Library of Congress catalog on October 17, 2018, for books about "planetary science" yielded 129 results. Titles indicate that these books are either about science or the history of the science.

Studies have been conducted on public interest in space exploration (Aviation Week, 2014; Funk & Strauss, 2018); the history of public opinion about NASA (Billings, 2010, pp. 171–177); the history of public opinion about US human space flight (Launius, 2003); the impact of space exploration on public opinion, attitudes, and beliefs (Bainbridge, 2015); public understanding of science (National Science Board, 2018)[4]; the public value of science (Wilsdon et al., 2005); public engagement with science (Entradas and Bauer, 2016; Fogg-Rogers et al., 2015; Leshner, 2003) and with space exploration (Space Studies Board, 2011); and public support for space science and exploration (European Space Agency, 2005). However, little research has focused specifically on public engagement with planetary science in particular (Billings, 2019).

In this chapter, I will report on efforts I have played a part in to improve and expand communication and engagement with nonexpert audiences about space science and exploration. In going about my work, I have applied insights drawn from the literature on science communication, social studies of science, the rhetoric of science, public understanding of science, and critical and cultural studies. I will focus in detail on efforts for two NASA planetary science programs that have sponsored my work—astrobiology, which funds research into the origin, evolution, and distribution of life in the universe; and planetary defense, which funds projects that find, track, and characterize near-Earth asteroids and comets (known as near-Earth objects, or NEOs) and works on planning to defend Earth from possible NEO impacts. I will offer brief case studies of specific projects, identify some themes about public engagement that emerged from them, and describe some lessons learned during the course of the conduct of the projects.

## Public Engagement With Astrobiology

Astrobiology is the study or the origin, evolution, and distribution of life in the universe—a topic that is inherently interesting to many nonexperts, if one can gauge interest by how much time and space mass and social media give to it. The search for evidence of life elsewhere seems to have a permanent presence in the public discourse.

The NASA astrobiology program employs a communication strategy aimed at promoting the widest possible dissemination of timely and useful information about scientific discoveries, technology development, new knowledge, and greater understanding produced by the program. Much of my work with NASA's astrobiology program over the years has focused on broadening and balancing the public discourse about the search for life in the universe. Media coverage of the topic tends toward wild-eyed optimism and outright hyperbole—especially in news headlines:

---

[4] A peer-reviewed journal called *Public Understanding of Science* has been publishing research on the subject since 1992.

Finding life on Mars is imminent! Any day now, we will make contact with extra-terrestrial intelligence! There must be life on Europa, and Enceladus, and we have to go there now! With thousands of extrasolar planets confirmed, soon we will find life on planets beyond our solar system! A persistent challenge is establishing that NASA is searching for evidence of microbial life in the solar system, not intelligent life in the universe.[5]

I will report on three projects I worked on for the NASA astrobiology program: FameLab USA, conceived and executed as a science communication initiative that also contributed to broader public engagement with astrobiology; Ask an Astrobiologist (AAA), an online public information service that adapted to the rise of social media; and a multifaceted effort to engage with scholars in the social sciences and humanities.

## Case Study: FameLab USA

FameLab is an international science communication competition organized by the UK-based Cheltenham Festivals and targeted to early-career scientists.[6] The application process is competitive, and the number of contestants in a competition is limited. Contestants are asked to make a 3-min presentation on a science concept using no slides, charts, or props beyond items they can carry on stage. A panel of judges provides constructive feedback on these presentations. Then the contestants are provided with several hours of training in how to construct a compelling narrative. Following training, contestants make another 3-min presentation, and the judges provide a critique. Final presentations are made before a public audience.

After learning about FameLab in reviewing a paper by George Zarkadakis (2010), a competitor trainer and head judge in Greece's competition, I found that it had spread across Europe and Asia but had not yet arrived in North America. In 2011, I briefed one of my NASA funders, senior scientist for astrobiology Mary Voytek, and her colleague Michael New, NASA discipline scientist for astrobiology, on FameLab, suggesting, "Wouldn't it be great if we could do something like this?" Voytek and New took up my suggestion, figured out how to fund it and who to put to work on it, and shepherded a FameLab USA[7] proposal through NASA to approval. The aim of this project was not to train participants to "sell" science but to train them in how to explain science to nonexpert audiences.

I served as a judge in FameLab USA's first year, 2011, and I was impressed at both the talent we saw for communicating across expert boundaries and also the commitment of many participants to embracing public engagement as part of their

---

[5] I credit this phenomenon to standard journalistic practices aimed at grabbing attention and also to a handful of popular "talking-head" scientists who are willing to make attention-grabbing claims that are not supported by data.

[6] https://www.cheltenhamfestivals.com/science/famelab/famelab-international-2/.

[7] https://famelab.arc.nasa.gov.

jobs.[8] The winner of the first year's competition, Brendan Mullan, was accepted to participate in FameLab International's global final competition. Mullan is now an assistant professor of astronomy at Point Park University in Pittsburgh, Pennsylvania. He characterizes himself as an "astrophysicist, science communicator, and education program developer and director."[9] The 2012 FameLab USA winner, planetary scientist Aomawa Shields, was invited to be a TED Fellow in 2015.[10] TED Fellows "participate in professional coaching and mentoring and work with a public relations expert dedicated to sharing their latest projects with the world."[11] Shields has founded an organization called Rising Stargirls, which is dedicated to encouraging girls of all colors and backgrounds to explore and discover the universe using theater, writing, and visual art.[12]

Over the course of the FameLab USA project (2011-16), roughly 250 early-career scientists from across the United States and multiple disciplines participated in the competition.[13] We were pleased that our FameLab recruitment efforts yielded a field of contestants who were diverse in gender, race, and ethnicity as well as in discipline. In December 2012, Voytek asked astrobiologist and FameLab alumna Heather Graham and me to work on organizing a FameLab Astrobiology Communication Network, starting by inviting everyone who competed in FameLab USA 2012 to join. SAGANet—"Social Action for a Grassroots Astrobiology Network"— an organization founded in 2012 by a group of early-career astrobiologists "passionate about public engagement in science"[14]—has taken over the job of sustaining this network. (The NASA astrobiology program provides some support to the organization.) Though I have not done any sort of formal survey, I have noted in the years since the start of FameLab USA that many FameLab participants have continued their work in public engagement with astrobiology—a particular challenge for early-career astrobiologists who are completing postdoctoral research projects and looking for jobs.

The long-term value of FameLab USA as a means of engaging public audiences with science would be difficult to assess. Such an assessment would require tracking the career paths of participants and their public engagement activities. As to short-term effects, FameLab USA staff surveyed participants before and after their competitions to assess whether they believed they had benefited from participation.

---

[8] FameLab USA final presentations are archived at: https://www.youtube.com/user/FameLabUSA.

[9] https://www.pointpark.edu/Academics/Schools/SchoolofArtsandSciences/Departments/NaturalSciencesandEngineeringTechnology/NaturalSciEngTechFacStaff/BrendanMullan. Accessed 10 December 2018.

[10] TED—Technology, Entertainment, and Design—is an organization dedicated to "spreading ideas." See: https://www.ted.com/speakers/aomawa_shields. Accessed 24 January 2019.

[11] https://www.ted.com/about/programs-initiatives/ted-fellows-program. Accessed 24 January 2019.

[12] http://www.risingstargirls.org. Accessed 24 January 2019.

[13] https://famelab.arc.nasa.gov/media/FameLabUSA-RetrospectiveReport-June2016.pdf. Accessed 10 December 2018.

[14] http://saganet.org/page/about. Accessed 10 December 2018.

Survey results showed that participants reported increased confidence and a higher likelihood of identifying as "science communicators."[15]

One theme that emerged from this project is that science is inherently interesting. With a bit of coaching, FameLab USA participants were able to construct engaging talks about current science, One lesson learned is that many early-career scientists are interested in participating in public engagement activities but may lack encouragement and support to do so. NASA's provision of encouragement and support through FameLab USA provided a boost in confidence.

## Case Study: Ask an Astrobiologist

AAA was a NASA-sponsored online service conceived to answer people's questions about astrobiology. Over a decade, it evolved into something else, and ultimately the Q&A format became outdated.

On October 29, 2001, the NASA Astrobiology Institute (NAI), an element of NASA's astrobiology program based at NASA's Ames Research Center, initiated AAA. For the first couple of years, various members of the astrobiology community volunteered to answer questions for AAA.

In February 2003, NASA Ames planetary scientist David Morrison, then the NAI's senior scientist, took over answering all questions for AAA. In 2008, the NASA astrobiology program established a web presence for the entire program, encompassing the NAI's website. With this development, AAA became a service of the astrobiology program, with NASA headquarters oversight. As a consultant to the astrobiology program on communication issues, I began monitoring content on AAA, and I grew increasingly concerned about the appropriateness and utility of the service.

In my review of 7 years worth of AAA content, I found that about 9 out of 10 questions (with answers) posted were not relevant to astrobiology. About half of those were about science other than astrobiology, and the other half fell into the category of what I would call pseudoscience and conspiracy theory (do UFOs exist, is the government hiding alien corpses, was the Moon landing a hoax, is the world ending, etc.).

Judging by the persistent submission of questions about the same fringe-y topics—in particular, questions about government conspiracies—it appeared that Morrison's attempts to debunk peoples' pseudoscientific and conspiratorial beliefs by countering with his own knowledge and points of view were not working—at least not on AAA.

Here is a typical example of an exchange with an (anonymous) individual on AAA (Morrison, 2012):

Q: Why are the US government building bunkers to house the elite in case of a global disaster? Also I heard about two suns in China. Is one of them Nibiru?

---

[15] See foot note 14.

A: These two questions both illustrate the Big Lie, which can happen when a story goes viral on the Internet. Both are old hoaxes that just will not go away. The government bunker rumor has two sources. First is the well-known fact that the US government built many deep air raid shelters during the early stages of the Cold War, including the underground complex at Greenbrier, West Virginia, which was supposed to provide protection for the President and his staff, and the Cheyenne Mountain complex in Colorado, which is still used by the Strategic Air Command. Second is an episode on bunkers in Jesse Ventura's TV series Conspiracy Theory. Unfortunately, many people do not realize this is an entertainment show, not journalism—something that is given away by the title, since the phrase "conspiracy theory" implies an interpretation that is not based on facts. But even his episode about shelters is all innuendo; Ventura does not actually interview anyone who claims to have seen one of the phantom "government bunkers."

The source of the "two suns in China" report is even less well understood by the public. If you watch the original news video with a friend who speaks Chinese, you will see that this is about a single photo, broadcast by a local TV station on a small island near Taiwan. It is based on a still picture (not a video), apparently rephotographed with a handheld video camera. As far as I can tell, the "two suns" were not seen elsewhere in Taiwan, let alone across China. And obviously if this phenomenon were real, billions of people everywhere could have verified it just by looking up at the Sun.

In discussing the state of AAA with my colleagues at NASA headquarters, we reached a consensus that AAA should be retired, revamped, rebranded, or replaced. At this point, I was asked to oversee a revamping of AAA. We asked early-career astrobiologist Heather Graham, a FameLab USA alumna, to assemble a team of AAA volunteers from among other FameLab USA alumni. We asked astrobiologist Sanjoy Som to assemble a team of AAA volunteers from among members of SAGANet.

The volunteers and I had a couple of productive conference calls to determine how to proceed with AAA. We agreed that the purpose of AAA was to inform public audiences about astrobiology. Some of the volunteers raised questions about the utility and effectiveness of the online Q&A format, given the proliferation of options for online engagement. But we agreed to try it out once more.

I would serve as "executive editor," and Graham and Som would manage their respective teams. While we agreed that AAA did not have to be devoted exclusively to astrobiology research supported by NASA, we agreed the service would focus on responding to queries about astrobiology. We would refer questioners to other reliable sources of information about other legitimate science topics. We also agreed that not all Q&As would be posted on the website.

Here are some examples of questions submitted to AAA over part of the period during which we were planning the AAA revamp. They are typical, rather than atypical, of submissions. Typos and misspellings are in the original. I do not have a record of whether or how I responded to some of these queries:

Q: "Please can you answer my question I am so scared about dying its my worse fear please could you reasure me that were not gonna die anytime soon with the galaxy moving apart please please help me." (April 2013)

(We received the same question from the same email address once a month for more than a year, at which point we blocked the email address from the service.)

Q: "Does the universe as a whole care about my friend Lara not being ill anymore?" (July 10, 2013)

My answer: "NASA's Astrobiology Program sponsors research into the origin, evolution, and distribution of life in the universe. Your question falls outside the purview of astrobiology, so we will not be able to answer it."

Q: "So, if astrobiology can't help me with this question, is there anyone who can?" (July 10, 2013)

(I did not respond to the second query.)

Q: "today in history channel a series are coming named ancient alien is these possible that in our past our ancestors encountered any alien, unfortunately they think they where gods?" (August 12, 2013)

A (I responded but did not post this Q&A to the website): "Scientists have not yet found any indisputable evidence, on Earth or in space, of extraterrestrial intelligent life, past or present. While some individuals may have claimed they have found evidence, none of these claims has been validated by the scientific community."

Q: "What i basically ask myself everyday is: is there any known form life elsewhere than on earth? In my opinion, yes there is, but is there any proof ? Because in africa, the Dogon tribe as clearly the answer but who to trust ? Is Xylanthia reachable by any known way. Will we ever try to communicate with the extra terrestrial life that is supposed to live there?" (November 12, 2013)

Q: Regarding Area 51? Groom Lake (aka Area 51) has started off as an airbase during the 1920s. During the Cold War, a U-2 Spy Plane was shot downed. Thus ended the era of the U-2 Reconisance missions. Afterward another project launched by the CIA would emerge, known as "Project Oxcart."

As a refresher, "Project Oxcart" is an operation that commenced in California. The project introduced a new type of military aircraft known as the SR-71. The SR-71, unlike the U-2 spy plane is capable of flying at higher altitudes, and higher speeds, and even has the "jet engine."

My real question is, why has the US kept "projects and military aircraft" a secret from the public? The SR-71 is very notorious, and it has been somewhat leaked to the public during the late 1970s-1980s. Was the United States really afraid of Russians that bad? Has the "Alien & UFO" hype leaked the existence of Area 51?" (February 5, 2014)

Here are a few of the questions I sent directly to a trash folder, without responding:

"what is god" (April 23, 2013)

"Why am I freezing? Why are scientists so freeking stupid these days? Who created the universe? Is the universe decaying since its creation? Who is in charge? God can answer all of these questions and YOU can not. Oprah" (January 24, 2014)

"I believe I could possibly be LUCA [the last universal common ancestor], I would like you to prove me wrong (for my own benefit) through a phone call [number deleted]. Any time possible. Any effort towards this phone call being made would be greatly appreciated. The second I am proven wrong, I will stop bugging you, as science requires proof." (January 12, 2014)

"i think i might be the cause of dark energy im in melbourne. I got sexually abused in 1993 that might be the cause if possible call me on [number deleted]." (July 18, 2013)

In December 2013, I conducted a webinar for AAA volunteers on challenges in communicating the complexities of astrobiology research. Then volunteers got to work. After a year or so of operations and a few conference calls, volunteers came to agreement that AAA was too time-consuming, the benefits could not be assessed, and the online Q&A format was outdated. Over the next several months, the SAGANet network, which was already hosting an online "Talk to an Astrobiologist" service independently, worked on a plan to take over and revamp AAA. SAGANet's "core feature" is now a livestreamed video series called AAA, produced with support from NASA and featuring live chatting by text or Twitter. The livestreamed events are archived on YouTube. SAGANet also offers an "ask an expert" service online: "We want to keep the conversation going 24/7 all around the world." SAGANet's take-over and revamp of AAA is an experiment in progress to expand public engagement in science.

As with FameLab USA, a theme that emerged from this project is that many early-career scientists are eager to participate in public engagement. One lesson I gleaned from my experience with AAA is that it would be virtually impossible to assess the benefits of the online Q&A AAA as a public engagement tool, as NASA is prohibited from collecting any information on participants, beyond email addresses. Another lesson learned is that the online Q&A format is not an effective means of responding to conspiracy theorists. Research has shown that people who fixate on conspiracy theories tend to reject "all scientific propositions tested …. The resistance of conspiracist ideation to contrary evidence renders its prominence in the rejection of science particularly troubling, because providing additional scientific information may only amplify the rejection of such evidence, rather than foster its acceptance' (Lewandowsky et al., 2013).[16] From where I stood, this is what appeared to be happening during AAA's earlier years.[17]

---

[16] Also see Lewandowsky, S., Gignac, G., & Oberhauer, K. (2013). NASA faked the Moon landing, therefore (climate) science is a hoax. *Psychological Science* 24(5), 622–633, https://doi.org/10.1177/0956797612457686; van Prooijen, J-W., Krouwel, A.P.M., and Pollet, T.V. (2015). Political extremism predicts belief in conspiracy theories, *Social Psychological and Personality Science* 6(5), 570–578, https://doi.org/10.1177/1948550614567356.

[17] Lewandowsky et al. (2013) suggest an alternative strategy to debunking: "conspiracist misconceptions of scientific issues are best met by indirect means, such as affirmation of the competence and character of proponents of conspiracy theories, or affirmation of other beliefs they hold dearly …. Such self affirmation is known to facilitate the dislodging of [conspiracy-oriented] attitudes … " (n.p.).

## Case Study: Broadening Engagement With the Social Sciences and Humanities

The activity I will describe here involves several projects intended to kick-start an ongoing engagement of the natural science of astrobiology with the social scientific and humanistic study of astrobiology. NASA's astrobiology program addresses questions that are multidisciplinary and require an interdisciplinary, even transdisciplinary approach to answer. Astrobiology also touches on issues in the humanities and social sciences, which focus especially on the future of life—for example, the question of how the discovery of extraterrestrial life might affect human cultures.

Historically, projects addressing "societal implications of astrobiology" in the United States have been sparse, sporadic, and disconnected. Until recently, questions relating to how the discovery of extraterrestrial life might affect "society" have been addressed by a small community of researchers largely involved with the search for evidence of extraterrestrial intelligence (SETI) and thus focused primarily on possible responses to the discovery of extraterrestrial intelligent life. Though members of the SETI community have long argued that contact with extraterrestrial intelligence will be a world-changing event, there is no evidence that this would be the case. In addition to a SETI-centric focus, past efforts to address "societal implications" also have been largely Western-centric—mostly US-centric. These efforts have reinforced the practice of cultural hegemony—the domination of a diverse culture or cultures by a ruling class—in this case, the domination of a myriad of wildly diverse global cultures by Western scientific, scientistic, culture, whose foundation is the Western scientific worldview.[18]

The field of astrobiology today is focused primarily on the search for evidence of past of present microbial life in the solar system, and thus discussion of ethical, philosophical, theological, and legal issues relating to astrobiology is broadening, and focusing, accordingly. In the late 1990s, the NASA astrobiology program decided to take some steps to broaden the scholarly and public dialogue on the search for extraterrestrial life, as NASA was facing growing scientific, political, and public interest in the subject in the wake of the publication of a paper in *Science* claiming a possible discovery of fossilized microbial life in a martian meteorite fragment (McKay et al., 1996). The program initiated several efforts to engage with the broader scientific community, scholars in the social sciences and humanities, and nonexpert audiences with astrobiology.

---

[18] To its credit, the Breakthrough Listen Initiative, a privately funded effort to search for evidence of extraterrestrial intelligent life, "now seeks to meaningfully engage with scholars who think with and around SETI as a way to socially, historically, and philosophically analyze our own science search. [The] main question is: What are we missing?" To this end, the Initiative held a workshop in April 2018 to engage with scholars of diverse backgrounds, disciplines, and theoretical perspectives on issues relating to "making contact." I participated in this workshop, and discussion was fascinating. See https://makingcontact2018.com/statement/.

First, the astrobiology program cosponsored a series of workshops in 2003-04 on the philosophical, ethical, and theological implications of astrobiology. The workshops were organized by the American Association for the Advancement of Science's Dialogue on Science, Ethics, and Religion program (Workshop Report, 2007).

In 2012, the NASA astrobiology program and the Kluge Center of the Library of Congress established a jointly endowed chair, the Baruch S. Blumberg NASA/Library of Congress Chair in Astrobiology. The Blumberg Chair—a project conceived by Nobel laureate Barry Blumberg, who was the first director of the NASA Astrobiology Institute—was created to support scholars interested in the intersection of the sciences and humanities. The chair creates an opportunity to study the range and complexity of issues related to how life begins and evolves and to examine the philosophical, religious, ethical, legal, cultural, and other concerns arising from scientific research on the origin, evolution and nature of life. Blumberg Chairs are expected to engage with members and staff of Congress. Chairs thus far have included astronomers, a planetary scientist, historians, and a humanities professor. The chairs are required to hold at least one public event during their tenure, most of which have been webcast. Chairs produce books or other publications about their work at the Kluge Center.

For 2014-15, the Blumberg Chair program did not select a chair. Instead, the program sponsored a series of interdisciplinary dialogues examining how astrobiology shapes and is shaped by religion and notions of creation, human self-understanding, history, culture, and art. Researchers who participated in these dialogues, led by philosophy professor Derek Malone-France of The George Washington University (GWU) and astrobiologist John Baross of the University of Washington, ranged from natural scientists to scholars of Buddhism, Christianity, Islam, and Judaism to STS (science, technology, and society) researchers, philosophers, historians, ethicists, and professors of English, theater, and rhetoric. Each of the closed dialogues was followed by a public program, which was webcast and archived online. Many of the participants remain engaged in public dialogue about astrobiology, in their classrooms, through publications, and in public talks.

One product of these dialogues is the organization of a scholarly network called the Society for Social and Conceptual Issues in Astrobiology (SSoCIA), cofounded by dialogue participant Kelly Smith, a philosophy professor at Clemson University. SSoCIA held its first workshop in Greenville, South Carolina, in 2015, attracting about 30 scholars. The network held a second workshop in Reno, Nevada, in 2017, attracting about 100 scholars. A third workshop, planned for spring 2020 at the University of Mississippi, had to be postponed due to the COVID pandemic. However, an online workshop was held in December 2020, involving 40 participants. SSoCIA is intent on bringing together a community of scholars who are not only diverse in discipline but also in gender, age, and other demographic factors.

The astrobiology program also supported the Center of Theological Inquiry's (CTI's) 2015-16 study-in-residence project, "Inquiry on the Societal Implications of Astrobiology," and the 2015-16 NASA Astrobiology Debates project, both intended to engage broader audiences with astrobiology.

CTI is an independent academic institution for interdisciplinary research on global concerns with an international visiting scholar program. (It is not a religious institution.) Questions that guided CTI's astrobiology inquiry, which involved 12 scholars in residence for a year, were as follows:

- If there are many different forms of life, known and unknown to us, what does it mean to be "alive"?
- How would art and literature depict life as we know it against this background of other possibilities?
- To what extent do our moral relations depend on the biology we share with other persons and other life?
- With all these unanswered questions about life in the universe, how do we organize ourselves to investigate the possibilities?

CTI obtained a separate grant from the Templeton Foundation for an outreach program to share the results of this inquiry with the broader academic community and public audiences during 2016-17.

The 2015-16 NASA Astrobiology Debates[19] was a year-long academic project, organized by faculty at GWU, for university and secondary education students involving in-person and online debate tournaments, speech competitions, public exhibition debates, topic-expert panels for student audiences, and student-conducted topic interviews with a cross-disciplinary group of subject matter experts. The aim of the project was to stimulate student, teacher, and school research and dialogue on astrobiology in preparing for these events and at the events themselves. The project engaged universities, subject matter experts, and elite college debaters to help develop content for an instructional website for secondary-school students who would be participating in the debates.

The initial goal of this project was to engage over 500 students in debate competitions. By spring 2016, the project's managers at GWU in Washington, DC, estimated that they would be involving over 2000 students during the 2015-16 academic year. A regional university debate competition held at GWU in 2016 attracted debaters from across the United States as well as Japan and France. Some of the debate competitions were live-streamed for parents, students, and other interested people who could not attend the events in person. Video records of these events were posted to the NASA Astrobiology Debates website to promote further public engagement.

NASA's most recent astrobiology roadmap, the 2015 Astrobiology Strategy, does not specifically identify goals, objectives, and questions relating to social, cultural, ethical, and theological issues arising in the study of the origins of life and the search for evidence of extraterrestrial life, because the community had embraced this endeavor as part of its ongoing work. However, in an appendix to the strategy, Lucas Mix and Connie Bertka—both with degrees in natural science and theology—identified a range of topics in the humanities and social sciences that could

---

[19] http://www.nasadebates.org. Accessed 20 December 2018.

"contribute to the central goals of astrobiology" and noted, "Encouragement of independent work in the humanities and social sciences on these topics will aid astrobiology immensely" (Hays, 2015, p. 160).

While I am not aware of any efforts to assess the long-term effects of these unconventional public engagement projects, from where I stand, at least in the short term, they appear to have been successful. A theme that emerged from these activities is that there is considerable interest among social scientists and humanistic scholars in engaging with scientists on questions of common interest. NASA's mandate does not include providing long-term funding for activities in the social sciences and humanities, and so these initiatives were conceived as short-term efforts to broaden engagement with the social sciences and humanities. A lesson learned is that without a source of funding, interested scholars are limited in their ability to engage in dialogue with the astrobiology community. The organization of SSoCIA is a positive outcome. I hope that this nascent community of scholars will continue to engage with astrobiology through interactions with colleagues, students, and public audiences as well.

## Public Engagement With Planetary Defense

I began working as a consultant to NASA's near-Earth object (NEO)[20] observations program on communication issues in 2012. (In January 2016, NASA established a Planetary Defense Coordination Office [PDCO], and the NEO observations program became its core element.) Our primary goal was then, and is now, to inform a wide variety of nonexpert audiences—especially decision makers and the media—about the research sponsored by the program—finding, tracking, and characterizing NEOs and determining whether any pose a risk of impact with Earth. As with astrobiology, NEO observations and planetary defense are too often misrepresented in mass and social media, with stories and headlines emphasizing the prospect of the end of the world by asteroid impact. Here, I will describe efforts to engage with citizens, policy-makers and decision-makers, and the media.

In 2010, the NEO observations program was handed a surprise: in April of that year, President Obama announced a new goal for NASA: a human mission to an asteroid. NASA consequently established an "Asteroid Initiative" incorporating planning for an asteroid retrieval mission (later named the Asteroid Redirect Mission) and a human mission to an asteroid, plus an "Asteroid Grand Challenge" (AGC), conceived as a large-scale, problem-solving public engagement effort.[21] The NEO observations program would provide data and expertise to the Asteroid Initiative.

---

[20] NASA defines NEOs as asteroids and comets whose orbits periodically bring them within approximately 1.3 Astronomical Units (AU) of the Sun. This implies that they can come within 0.3 AU—about 30 million miles, or 50 million kilometers—of Earth's orbit.

[21] https://www.nasa.gov/content/asteroid-grand-challenge. Accessed 26 February 2019.

## Case Study: The Asteroid Citizen Forums Project

Arizona State University's Consortium for Science Policy and Outcomes (CSPO) is a founder and member of Expert and Citizen Assessment of Science and Technology (ECAST), a consortium of universities, science centers, citizen science platforms, and nonpartisan policy think tanks. In 2013, NASA issued a request for proposals for engaging public audiences in the Asteroid Initiative. At an Asteroid Initiative Ideas Synthesis Workshop held in Houston in November 2013, CSPO's David Guston pitched an idea for a citizen engagement project. In April 2014, NASA entered a cooperative agreement with ECAST to conduct this project (Tomblin et al., 2015). Though I was not directly involved in this project, I observed its development and attended a briefing at NASA headquarters on its results.

ECAST organized two all-day citizen's forums, held in Boston, Massachusetts, and Phoenix, Arizona, in November 2014. The aim of these events was to engage nonexpert citizens in participatory technology assessment, providing them with an opportunity to "learn about, discuss and share their views on … planetary defense, NASA's Asteroid Redirect Mission (ARM) and scenarios for human and robotic missions to Mars." ECAST made a point of assembling participants (88 in Boston, 98 in Phoenix) who were diverse in age, gender, ethnicity, and other demographic factors.

This project had two primary goals: "to develop and apply a participatory technology assessment that elicited nuanced information from a diverse group of citizens whose insights would not otherwise be available to decision-makers" and "through informed, structured feedback from citizens in multiple locations … to provide public views of the Asteroid Initiative as input into NASA's decision-making process." Questions that guided discussion at these forums were: "What would an effective detection system that could improve humans' ability to protect Earth look like? If we had the capability to find all of the asteroids with the potential to cross Earth's path, what options might be available to address a detected threat? Thinking further into the future, if we develop the capacity to redirect an asteroid or piece of an asteroid then send astronauts to study it, what might those capabilities mean for future space exploration? Could they enable or support the ultimate ambition of a crewed mission to Mars?" (Tomblin et al., p. 4).

ECAST staff facilitated and recorded discussions, and NASA subject matter experts were on hand to answer questions. As NASA was deliberating on two options for an asteroid-retrieval mission, forum participants were briefed on both and asked which option they favored. They by and large preferred the option of retrieving a boulder from an asteroid and bringing it into lunar orbit for study versus retrieving a whole asteroid. (In 2015, NASA chose the boulder-retrieval option.) Participants agreed that NASA should move forward with space-based observation of asteroids rather than maintaining or expanding its current ground-based observation network. (The PDCO supports moving forward with a mission proposal called the NEO Surveillance Mission (NEOSM)—a space-based NEO survey telescope. However, the PDCO's budget is insufficient to develop NEOSM.) As to leadership in planetary defense, participants preferred that NASA work in partnership with other nations rather than pursuing planetary defense unilaterally. (The PDCO is working with international partners on planetary defense.)

ECAST took steps to assess outcomes of these forums. "Surveys helped determine participants' attitudes toward and knowledge of asteroid and space exploration before and after the forums and measured participant satisfaction with the experience" (Tomblin, et al., p. 8). Postforum survey results showed that "participants were highly satisfied with their experience at both sites. Pre- and postforum surveys indicated that participants greatly increased their interest and knowledge in NASA's Asteroid Initiative and plans for space exploration." Participants' attitudes about different planetary defense scenarios … also shifted, for example toward more agreement for the need for government support for space activities and the importance of international collaboration" (Tomblin et al., 2015, p. 18).

A report issued by the Government Accountability Office in October 2016 cited the Asteroid Initiative project as an example of "practices that promote the effective implementation of open innovation strategies" (Lloyd, 2016).[22]

My NASA colleagues who were involved with this project were impressed by the results (as was I) and, I believe, now better understand the value of participatory technology assessment in government decision-making. It remains to be seen whether the Biden administration will be more open to participatory technology assessment than the Trump administration was. But should NASA choose to pursue such activities in the future, the ECAST project offers a good model to follow—well conceived, well designed, well staffed, and well documented. This sort of public engagement is the democratic thing to do. Lessons learned here were that citizens are willing (perhaps even eager) to participate in government decision-making, proper expertise is necessary to conduct meaningful engagement, and thorough documentation of the engagement project is useful to the sponsoring agency.

## Case Study: Asteroid Day

Another public engagement project I worked on with PDCO staff was cooperation with "Asteroid Day." Asteroid Day[23] is both an annual event, held on June 30, and an organization, cofounded in 2014 by Queen guitarist Brian May, ex-NASA astronaut Ed Lu, filmmaker Grig Richters, and B612 Foundation CEO Danica Remy. Asteroid Day organizers describe the annual event as "a global awareness campaign where people from around the world come together to learn about asteroids, the impact hazard they may pose, and what we can do to protect our planet, families, communities, and future generations from asteroid impacts."[24] Asteroid Day holds public events in Europe on June 30, and locally organized Asteroid Day events take place around the world.

---

[22] https://www.gao.gov/products/GAO-17-14. Accessed 17 December 2018.
[23] https://asteroidday.org. Accessed 4 January 2019.
[24] https://asteroidday.org/about/. Accessed 4 January 2019.

The first annual Asteroid Day took place on June 30, 2015. Richters approached NASA later that year about sponsoring future Asteroid Days. My NASA colleagues and I agreed that sponsorship would not be useful. We were concerned about the Asteroid Day organization's scare-mongering rhetoric. For example, on the Asteroid Day website, Brian May offers this statement: "The more we learn about asteroid impacts, the clearer it became that the human race has been living on borrowed time." And here is another, misleading, quote from the website: "More than 1M asteroids have the potential to impact Earth and through all the available telescopes worldwide, we have discovered only about one percent." We also were concerned about the organization's aggressive fundraising and publicity campaigning, and inattention to all the work already done in finding, tracking, and characterizing NEOs and planning for planetary defense.

We did engage with Richters about our concerns, and to his credit, in 2016, he did work on toning down Asteroid Day's rhetoric and adding more credible scientific information to the Asteroid Day website (though much of it now appears to have been removed). For Asteroid Day 2016, the PDCO provided content to the Asteroid Day organization for an "educational toolkit," a PowerPoint presentation on in-space NEO impact mitigation techniques, and an accurate narrative for a PowerPoint show about the history of asteroid impacts in the solar system. For Asteroid Day 2017, NASA provided an hour of original television broadcasting, produced by the Jet Propulsion Laboratory and including animations, recorded content, and live interviews.

While Asteroid Day's organizers claim credit for creating a "global grassroots campaign with tens of thousands of supporters across the globe," NASA has no way of gauging the impact of its contributions to Asteroid Day's public engagement efforts or the accuracy of Asteroid Day's claims about reach. The PDCO does not plan further engagement with Asteroid Day.

## Case Study: Engagement With the International Community

Following recommendations coming out of the United Nations' UNISPACE III conference in 1999, the Scientific and Technical Subcommittee of the UN Committee on the Peaceful Uses of Outer Space (COPUOS) established an "action team" to come up with a plan for coordinating international efforts to assess the risk of NEO impacts with Earth. The action team recommended forming an International Asteroid Warning Network (IAWN) to organize and coordinate member states' NEO observation efforts and a Space Mission Planning Advisory Group (SMPAG) to work on plans for deflection or mitigation.

NASA and the European Space Agency (ESA) took the lead in standing up IAWN and SMPAG, working in consultation with COPUOS and its parent organization, the UN's Office of Outer Space Affairs (OOSA). All of these organizations are a key means of sustaining engagement with the broader global community on cooperation in finding, tracking, and characterizing NEOs, assessing NEO impact risks, and developing technologies for deflection missions.

I attended the first meeting of the IAWN steering committee, in January 2014. I was tasked with organizing a panel discussion for this meeting on challenges in NEO science and risk communication. I chose panelists who would present a diversity of perspectives on the topic including disaster management, risk communication, and global reinsurance. This panel discussion led the IAWN steering committee to recommend holding a 2-day IAWN workshop on communication strategy and planning regarding NEO impact hazards. I was tasked with organizing this workshop, which was held in Broomfield, Colorado, in September 2014. In October 2015, ESA held a similar workshop in Frascati, Italy.

In February 2016, I was invited to participate in an "open forum" at a meeting of the Scientific and Technical Subcommittee of COPUOS in Vienna, Austria, on the status and activities of IAWN and SMPAG. My presentation was on IAWN communication planning. This forum drew a globally diverse audience, and since then, several more UN member states have joined IAWN and SMPAG. Again, it is difficult to gauge the effects, or benefits, of this sort of engagement effort, targeted to policy makers and decision makers. But I believe these efforts are worthwhile. A lesson I have learned, over and over again, including through this activity, is that a key to meaningful engagement with policy-makers and decision-makers is person-to-person, sustained contact.

## Case Study: Engagement With the Media

The community of NEO observers is well organized to report on so-called "close approaches"—predicted events in which asteroids pass our planet within a few lunar distances (1 LD=approximately 240,000 miles, or 384,000 kilometers) of Earth, but they are of particular interest when they pass within the orbit of Earth's Moon on their orbits around the Sun.[25] While these close approaches tend to be measured in tens of thousands to hundreds of thousands or even millions of miles and do not pose any risk of impact with Earth, media reports tend to report these occurrences as "close shaves" or "near misses." And media reports about asteroid impact hazards tend to emphasize the possibility of a catastrophic impact—"world-ending," "civilization-destroying," and so on. My NASA colleagues and I decided that we needed to engage directly with the media to inform them about NEO impact hazards. Our aim was to establish face-to-face contacts, and perhaps relationships, with journalists who are reporting on NEO impact risks.

Our first media engagement project was a seminar for science writers, held in conjunction with the annual meeting of the American Astronomical Society on January 9, 2014, in National Harbor, Maryland: "Everything you've always wanted to ask about near-Earth objects: what we know, what we don't know, what we need to know." Our next media engagement project was a workshop for broadcast meteorologists held at NASA headquarters in Washington, DC, on March 16, 2018. We

---

[25] Some passes of larger NEOs close to the Earth-Moon system but not between the two bodies are also called close approaches.

targeted broadcast meteorologists because they frequently report on NEO close approaches and other cosmic phenomena. We requested, and received, feedback from participants in this workshop, and we used that feedback to improve our next media engagement activity: a workshop for science writers held December 9, 2018, in Washington, DC, in conjunction with the annual fall meeting of the American Geophysical Union.

As participants in all three of these workshops indicated that they found them useful, the PDCO intends to conduct similar workshops in conjunction with other scientific conferences. PDCO staff are also considering the creation of a workshop "package" that members of the NEO science community can use for engagement with public audiences in schools, libraries, or other community centers. Again, it would be difficult to gauge the benefits of these activities, as it would require long-term follow-up with participants.

I continue to keep an eye on media coverage of NEO observations, NEO close approaches, NEO impact risks, and plans for planetary defense. Though I will not be engaging in any sort of formal analysis, I do believe that the more time and effort the PDCO team puts into propagating clear, concise, correct, and timely information about these topics, the better media coverage will be. A lesson learned here is that most journalists welcome access to reliable and authoritative information and that the burden is on us, the scientists, to reach out to them. We also have learned that, no matter what we do, we will not be able to dissuade the British tabloids from propagating melodramatic headlines, but we can counter their effects by cultivating trust-based relationships with other members of the media.

## Conclusions, Lessons Learned, and Recommendations

It would be difficult, if not impossible, to assess the value of investments in public engagement with planetary science activities. As to the public engagement activities I have worked on with NASA planetary science programs, most have been short-term rather than sustained. They have provided an opportunity for NASA scientists to learn more about what people know, do not know, and want to know. They have provided an opportunity for citizens to make contact with legitimate scientists. They have provided an opportunity for NASA scientists to properly frame issues that the media (and others) often misrepresent. It is reasonable to believe that these efforts have made a contribution to building public trust in science and in government. I believe that these efforts have been worthwhile. But they are not enough.

I have observed that the aims of planetary scientists in public engagement and the aims of NASA in public engagement are not necessarily the same. I do believe that scientists whose work is publicly funded (including myself) have an obligation to be able to explain the public value of the research they do. I agree with Peterman et al. (2017): "There is no "best" approach for public engagement …; instead, scientists must reflect on their goals for engagement and determine which approaches might be most appropriate for facilitating conversation and achieving their desired outcomes" (p. 783). That said, as a government agency, NASA is driven by politics and policy. Space politics and policy tend to reflect the desires of the military-industrial complex, not the desires of citizens.

AAAS's Center for Public Engagement with Science & Technology (Center, 2016) has developed a "theory of change for public engagement." This theory of change provides "a common framework, language, and research-based foundation for … professionals involved in public engagement with science activities, and to serve as a starting point to enable scientists, practitioners, and researchers to continually improve and develop collective understanding of effective practices in public engagement with science." It is based on a review of the literature that is relevant to public engagement. It would be useful for scientists who want to engage with public engagement to familiarize themselves with this body of literature, which is neatly summarized in AAAS's "theory of change" document.

Research has shown that public engagement with science efforts tends to engage people who are already engaged with science (Dialogue, 2018; Fogg-Rogers et al., 2015). The challenge is to reach, and engage with, people who are not interested or engaged, so that all citizens may be equipped to make informed decisions about science-related issues.

In closing, I would like to reemphasize that true public engagement involves listening and responding (Dialogue, 2018). The NASA-ECAST asteroid citizen forums are a good example of how to truly engage with citizens in decision-making. For decades, NASA has promoted its primary mission as sending people back to the Moon and on to Mars. It is doing so today. Yet there is little indication that NASA's Moon-Mars "visions" have had, or do have, broad public support. A Pew Research Center survey conducted in 2018 (Funk & Strauss, 2018) offered respondents nine priorities for NASA and asked them to rank them in order of priority. Respondents ranked monitoring Earth's climate and monitoring asteroids that pose a risk of impact with Earth as the top two priorities for NASA. Respondents ranked sending humans to the Moon and Mars last.

Based on my experience as well as my review of the literature on engagement, I have come to believe that public engagement with science may be most effective at the local level, where scientists can establish lasting relationships with citizens. For the past several years, I have spent an hour once or twice a week as a volunteer "scientist in the classroom" at local elementary schools. I have worked with classes in grades 2 through 5. I have learned a lot about communication and engagement through my meetings with students. I have learned that children are way smarter than you might think. I have learned that children never get enough science in the classroom, even at a science focus school. I also have learned that the challenge of engaging students with science is greater in a Title 1 school, where students often are not getting enough sleep or proper nutrition and may not have help with school work at home. And I have learned that, for all children, science needs to be relevant.

I have pursued other opportunities to interact with public audiences at the local level. I have given three "issues forum" presentations at my church, the Unitarian Universalist Church of Sarasota (Florida), on the ethics of colonizing Mars (2017), planetary defense (2018), and the U.S. Space Force (2020). They were well attended and well received, and I am now in continuing dialogue with a number of people who attended. I continue to look for new opportunities to engage with people about planetary science.

## Acknowledgments

Work on this chapter was funded by NASA Cooperative Agreement NNL09AA00A.

## References

Aviation Week Market Briefing. (June 26, 2014). *Public interest in space exploration*. http://aviationweek.com/site-files/aviationweek.com/files/uploads/2014/06/asd_06_26_2014_cht.pdf (Accessed 10 December 2018).

Bainbridge, W. S. (2015). The impact of space exploration on public opinion, attitudes, and beliefs. In S. J. Dick (Ed.), *Historical studies in the societal impact of spaceflight* (pp. 1–76). Washington, DC: National Aeronautics and Space Administration.

Billings, L. (2010). 50 years of NASA and the public: What NASA? What publics? In S. J. Dick (Ed.), *NASA's first 50 Years: Historical perspectives, NASA history division (NASA SP-2010-4704)* (pp. 151–182). Washington, DC: National Aeronautics and Space Administration.

Billings, L. (2019). *Public impact of planetary science, in press, the Oxford research encyclopedia of planetary science*. New York: Oxford University Press.

Center for Public Engagement with Science & Technology. (2016). *Theory of change for public engagement with science*. Washington, DC: American Association for the Advancement of Science. https://www.aaas.org/programs/center-public-engagement-science-and-technology/theory-change-public-engagement-science (Accessed 12 December 2018).

Dialogue on Science, Ethics and Religion and Center for Public Engagement with Science & Technology. (2018). *Scientists in civic life: Facilitating dialogue-based communication*. Washington, DC: American Association for the Advancement of Science. Available at https://www.aaas.org/sites/default/files/s3fs-public/content_files/Scientists%2520in%2520Civic%2520Life_FINAL%2520INTERACTIVE%2520082718.pdf.

Dick, S. J. (2014). Introduction: The impact of the hubble space telescope,. In R. D. Launius, & D. H. DeVorkin (Eds.), *Hubble's legacy: Reflections by those who dreamed it, built it, and observed the universe with it* (pp. 74–78). Washington, DC: Smithsonian Institution Scholarly Press.

Entradas, M., & Bauer, M. M. (2016). Mobilisation for public engagement: Benchmarking the practices of research institutes. *Public Understanding of Science, 26*(7), 771–788. https://doi.org/10.1177/0963662516633834 (Accessed 12 November 2018).

European Space Agency. (2005). *The impact of space activities on society. ESA BR-237*. http://www.esa.int/esapub/br/br237/br237.pdf (Accessed 10 January 2019).

Fogg-Rogers, L., Bay, J. L., Burgess, H., et al. (2015). Knowledge is power: A mixed-methods study exploring adult audience preferences for engagement and learning formats over 3 years of a health science festival. *Science Communication, 37*(4), 419–451.

Funk, C., & Strauss, M. (2018). *Majority of Americans believe it is essential that the U.S. remain a global leader in space*. Washington, DC: Pew Research Center. http://www. pewinternet.org/2018/06/06/majority-of-americans-believe-it-is-essential-that-the-u-s-remain-a-global-leader-in-space/ (Accessed 7 June 2018).

Hays, L. E. (Ed.). (2015). *NASA astrobiology strategy*. Washington, DC: National Aeronautics and Space Administration. https://astrobiology.nasa.gov/uploads/filer_public/01/28/01283266-e401-4dcb-8e05- 3918b21edb79/nasa_astrobiology_strategy_2015_151 008. pdf (Accessed 20 December 2018).

Launius, R. D. (2003). Public opinion polls and perceptions of U.S. human spaceflight. *Space Policy, 19*(3), 163−175. https://doi.org/10.1016/S0265-9646(03)00039-0 (Accessed 12 June 2018).

Leshner, A. (2003). Public engagement with science. *Science, 299*, 977.

Lewandowsky, S., Gignac, G. E., & Oberauer, K. (2013). The role of conspiracist ideation and worldviews in predicting rejection of science. *PLoS One, 10*(8). https://doi.org/10.1371/journal.pone.0134773 (Accessed 8 June 2018).

Lloyd, J. (2016). *ECAST featured in GAO open innovation report. Tempe, AZ: Consortium for Science Policy and Outcomes*. https://ecastnetwork.org/2016/10/18/ecast-featured-in-gao-open-innovation-report/ (Accessed 3 January 2019).

McKay, D. S., Gibson, E. K., Thomas-Keprta, K. L., Vali, H., Romanek, C., Clemett, S., Chillier, X., Maechling, C., & Zare, R. (1996). Search for past life on Mars: Possible relic biogenic activity in martian meteorite ALH84001. *Science, 273*(5277), 924−930. https://doi.org/10.1126/science.273.5277.924 (Accessed 15 December 2018).

Morrison, D. (2012). Ask an astrobiologist. *Skeptical Briefs, 22*(2). https://www.csicop.org/sb/show/question_what_is_the_history_behind_astrobiology (Accessed 15 December 2018).

National Science Board. (2018). Science and technology: Public attitudes and understanding. In *Science and engineering indicators 2018* (pp. 52−67). Washington, DC: National Science Board. https://www.nsf.gov/statistics/2018/nsb20181/assets/404/science-and-technology-public-attitudes-and-understanding.pdf (Accessed 8 December 2018).

Peterman, K., Evia, J. R., Cloyd, E., & Besley, J. (2017). Assessing public engagement outcomes by the use of an outcome expectations scale for scientists. *Science Communication, 39*(6), 782−797.

Space Studies Board, National Research Council of the National Academies. (2011). *Sharing the adventure with the public: The value and excitement of 'grand questions' of space science and exploration*. Washington, DC: National Academies Press.

Tomblin, D., Worthington, R., Gano, G., et al. (2015). *Informing NASA's asteroid initiative: A citizen's forum. Tempe, AZ: Expert and citizen assessment of science and technology*. https://www.nasa.gov/sites/default/files/atoms/files/ecast-informing-nasa-asteroid-initiative_tagged.pdf (Accessed 3 January 2019).

Wilsdon, J., Wynne, B., & Stilgoe, J. (2005). *The public value of science: How to ensure that science really matters*. London: Demos.

Workshop Report. (2007). *Philosophical, ethical and theological questions of astrobiology*. Washington, DC: American Association for the Advancement of Science.

Zarkadakis, G. (2010). FameLab: A talent competition for young scientists. *Science Communication, 32*(2), 281−287. https://doi.org/10.1177/1075547010368554

## Further Reading

Besley, J. C., Dudo, A. D., Yuan, S., & Ghannam, N. A. (2016). Qualitative interviews with science communication trainers about communication objectives and goals. *Science Communication, 38*(3), 356–381.

Besley, J. C., & Tanner, A. H. (2011). What science communication scholars think about training scientists to communicate. *Science Communication, 33*(2), 239–263.

# Amateur Astronomy: Engaging the Public in Astronomy Through Exploration, Outreach, and Research

Sanlyn R. Buxner, Michael T. Fitzgerald, Rachel M. Freed

## Introduction

No book on space science and public engagement would be complete without a chapter on amateur astronomy. Writing a comprehensive chapter on amateur astronomers is a seemingly impossible task, as there is no single way to characterize an "amateur astronomer." Additionally, there is not enough space in one chapter to truly unpack and describe the complexity and diversity of amateur astronomers in the United States, let alone across the globe.

Our interest in amateur astronomy is personal. Collectively, we represent a professional astronomer, education researchers, public engagement professionals, and amateur astronomers. Rachel has been an amateur astronomer highly engaged in education and public outreach for the past 20 years. She regularly sets up her telescopes around town to share the night sky and interesting celestial events with the public. She belongs to five astronomy clubs in the greater San Francisco Bay Area and speaks internationally about the astronomy research seminars she runs for high school and undergraduate students and instructors. Michael is a professional astronomer and education researcher who actively supports amateurs through encouragement and promotion of their work with students of all ages. He has a keen interest in providing supporting material, software, and guidance to people who otherwise would not be able to access astronomy directly. Sanlyn is an educational researcher, project evaluator, and public engagement and communication professional with specific interests in free-choice learning and astronomy, having worked in and around public astronomy programs for 25 years.

Amateur astronomers and amateur astronomy clubs are, at various points, both collaborators for professional scientists and education and outreach professionals as well as their target audience. In this chapter, we explore the three largest roles of amateur astronomers today: as knowledgeable members of the public who find personal enjoyment in learning and engaging in astronomy, as knowledge brokers who engage in public outreach, and as knowledge producers who play an important role in scientific research. We provide case examples that serve to illustrate the breadth and depth of the types of amateur engagement. At the end, we discuss how the field has evolved since the beginning of the COVID-19 worldwide pandemic. Many of the recent innovations may serve as permanent changes for amateur astronomy, building on the ever-evolving nature of online technology and ways that we connect to others.

A review of the literature in astronomy and engagement reveals that most is focused on professional scientists engaging within education in formal contexts and less so in public outreach. There is also literature that looks at informal educators in science museums and planetariums (Plummer & Small, 2013; Slater & Tatge, 2017) who are not professional astronomers but are professional astronomy educators. There is much research on how astronomy is taught in both undergraduate (Waller & Slater, 2011) and K-12 educational contexts (Pompea & Russo, 2020; Salimpour et al., 2020). Despite the large and relatively active population of amateur astronomers, there is little peer-reviewed literature on this topic. The lack of systematic study of amateur astronomers' interest, motivations, and activities forces us to rely on conference proceedings, books, and other reports, as well as testimonials of individuals collected for this chapter. Information for this chapter was collected by the authors through interviews, email correspondence, and targeted stories collected by the International Astronomical Union during the summer and fall of 2020. In total, this represents 53 individuals living in the United States, Australia, Japan, Tunisia, Portugal, Indonesia, Cameroon, Nicaragua, and Botswana. The wide range of perspectives came from those who are amateur astronomers and those who work with and support amateurs, including professional astronomers, telescope vendors, and others.

## Characterizing Amateur Astronomers

The number of people involved in the amateur astronomy community is huge. A ballpark range is between hundreds of thousands and millions of people around the world who would identify as an amateur astronomer. Amateur astronomy has had a long and complex history. The 19th century saw wealthy "Grand Amateurs" making observations at a similar level to the professionals. This largely gave way to what we saw in the United States and in other developed Western democracies during most of the 20th century: middle-class people, predominantly men aged 35–65, with leisure time to devote to small-aperture astronomy as a hobby and occasionally to contribute to research, particularly in variable-star astronomy (Kannappan, 2001). The military need for optical instruments during World War II had a significant effect on the development, accessibility, and affordability of optics and equipment, which led to telescope factories and thus increased the ability of amateurs to easily obtain telescopes (Cameron, 2010, p. 11795). Today, amateur astronomers in Western societies remain largely Caucasian, well educated, male, and with significant disposable income. A survey by the Night Sky Network, a US-based coalition of amateur astronomy clubs, reports that "the majority of respondents were over 50 years of age and 84% were male" (White & Prosper, 2014). A recent study of hobbyists, which included 879 amateur astronomers, found that 92% were Caucasian, 85% were male, and the mean and median age were both over 50 years of age (Corin et al., 2018). More than 50% of the respondents had college experience and 40% had either a master's degree or PhD. The researchers found a very low percentage of racial and ethnic minorities in their sample set, not just in astronomy but across all hobbies studied, from beekeeping to falconry to home brewing.

Amateur astronomers engage in astronomy on a continuum from consumer to user to producer. Past work has described different constructs to characterize the diversity of amateur astronomers; on a continuum from "recreational participants" consisting of observers, armchair astronomers and telescope makers to "scientific observers" (Williams, 2000). Percy (1998) uses a broader definition, "I prefer to define amateur astronomer extremely broadly. In this case, their education, knowledge, skills at instrumentation, computing, observing, teaching and other astronomical activities could be anything from zero to PhD level in astronomy or a related field"....[the key] is that they are volunteers." Cameron (2010, p. 11795) defines an amateur astronomer as "an intervening actor between the scientific elite and the public." Gada et al. (2000) conceptualize amateur astronomers' engagement through a pyramid. Those with casual interest are at the base of the pyramid. Above those are the novice astronomers who are getting started and more engaged in the hobby of astronomy. Next are experienced amateurs who have fundamental knowledge of astronomy and the night sky, and many are members of clubs. At the top of the pyramid, just before professional astronomers, are master astronomers, who they characterize as individuals who "have skills, equipment, and know-how to conduct their own research and are anxious to participate in scientific research projects" (p. 16). Storksdieck and Berendsen (2007) classify amateurs broadly into categories of individuals who participate in research, enthusiasts whose passion is to learn about astronomy, and those engaged in observation with telescopic equipment.

Regardless of how we place amateurs into groups, as Percy (1998) describes, "amateur astronomers are united by one characteristic—their interest and enthusiasm for astronomy. Considering that there are at least 10 times more amateurs than professionals, they are an ally which we should not ignore." For the sake of this chapter, we will discuss amateurs in three broad categories that take into consideration our original premise that today amateurs are both consumers and producers of knowledge and the roles they play.

- *Amateurs as independent explorers*—Intrinsic motivation is a key factor for many amateur astronomers. In this sense, we will talk about amateurs who consume information and who observe and take observations for their own satisfaction.
- *Amateurs as outreach agents*—Sharing knowledge and experience is important to many amateurs. They were once members of the public whose interest was sparked and now are interacting within, and beyond, their amateur communities.

While a professional scientist could provide a spark to a member of the general public, the professional scientist returns to their institution, maybe never to reengage with the same public. Amateurs have more time, and often more social links, to sustain connections within communities.

• *Amateurs as researchers*—Some amateurs are particularly interested in contributing to science and sharing and discussing their results throughout their social network as well as more formal publication networks. Included in this realm are the growing number of individuals who participate in astrophotography, photometry, and spectroscopy to contribute to research.

In the sections that follow, we describe the diversity of experiences of amateurs within each of the three categories and present examples of individuals, clubs, events, and other organizations who all participate in the amateur astronomy ecosystem to show the complexity and richness of engagement. Individuals often occupy multiple categories and move fluidly between them based on their current motivation, inspiration, and life circumstances.

## Amateurs as Independent Explorers

Amateurs are independent explorers in a variety of ways. On one side of the spectrum are armchair astronomers and avid knowledge seekers. They are inspired by some aspect of astronomy and continually seek knowledge. They report personal interest and motivation to know more and watch TV shows, go to lectures, attend star parties, and read magazines and online material as much as they can about astronomy. Many of these individuals do not own telescopes, nor do they have an interest in staying up late observing with their own equipment. They want to talk to others about astronomy to bolster their own knowledge. Learning may be done at home or be supported by clubs, lectures, outreach events, and other organizations. These interactions provide a social network of like-minded individuals who care deeply about the same topic that they do.

Among them are observers, individuals who actively look at the sky with some type of equipment that they own or borrow. Observers may engage in activities daily or just a couple days a month or year, often when there is a notable celestial event. Observers may engage with observing campaigns, such as "Messier Object Marathons" in which they are challenged to find all 110 Messier objects, deep sky galaxies, nebulae, and star clusters in the sky in a single night. Observers often invest their own resources in buying binoculars, telescopes, and even building backyard observatories, for the love of observing. Some observers make it a highly personal activity, while others make it the center of their social structure.

Imagers are a subset of observers interested in pushing the boundaries of what they can "see" in the night sky. Moving beyond what they can observe through an eyepiece, these individuals use cameras and other equipment to take images and collect data on objects, using astrophotography, spectroscopy, and photometry, all made possible through digital camera technology that makes taking images of the night sky widely accessible. Imagers engage in these activities for their own satisfaction as well as for sharing with other amateurs and the general public and sometimes for research. Such individuals convene in events such as the Texas Star Party, which attracts 400 visual observers and astrophotographers every year for a week under the dark Texas skies. Observers may also belong in the other categories of outreach agents and researchers further described in the following.

Some amateurs who enjoy exploring independently are *telescope makers*, a niche group of individuals who enjoy the thrill of making. With the ever-increasing supply of inexpensive, high-quality telescopes, and cameras, fewer individuals are left needing to make their own, but rather they engage in making telescopes for their own satisfaction. The Springfield Telescope Makers, commonly known by the name of the associated clubhouse and annual convention, Stellafane, started in 1923 (https://stellafane.org/). The group has about 125 members who are geographically dispersed around New England and have a shared passion for building telescopes. Full membership requires that a member has made a telescope objective lens. In addition to making telescopes, they provide classes for others and an annual conference. Another group designing and building telescopes is the Alt-Az Initiative (http://www.altazinitiative.org/), which has been meeting and collaborating for over a dozen years. One member, Don Peckham, has been approached by an international governmental research organization to collaborate on building innovative ground- and space-based telescopes based on his Tensegrity telescope (Skelton, 2013). Additionally, local telescope clubs and science centers across the United States still support interested telescope makers locally by providing hands-on workshops. Thus, while telescope makers act as independent explorers, they can also serve as outreach agents and collaborate to impact professional astronomy.

## Amateurs as Outreach Agents

Some amateurs serve as outreach agents who not only have a deep love of astronomy but also want to share their knowledge, passion, and enthusiasm with others. Some love to share observing experiences, while others love to share current science and exploration information and do not engage in observing. The latter is especially the case in developing countries among amateurs who do not own or have access to a telescope. Popular among amateurs who do outreach are those who do "sidewalk astronomy" in which individuals use telescopes to provide stargazing opportunities to passersby. Outreach-oriented amateurs' intentions are wide ranging, from conveying personal excitement to increasing meaningful experiences for others to teaching science. Most report satisfaction in inspiring excitement and awe in others, a common example being evoking the "ahhhhh" sound heard when people first see Saturn through a telescope.

Amateurs aid in bringing the excitement of astronomy to the public through a variety of organizations and projects. Popscope (https://www.popscope.org/), made up of volunteers who love astronomy and expanded from its origins in Ottawa, Canada, now includes hundreds of public events on five different continents. It uses a flash mob approach to bring telescopes to urban centers and share astronomy with many communities who are often left out of or cannot attend other telescope observing events. Access to telescopes is also enabled by the Library Telescope project (https://www.librarytelescope.org/). Started by Marc Stowbridge and the New Hampshire Astronomical Society, the initiative has expanded to over 40 states and libraries of all sizes. Through the project, an Orion StarBlast 4.5″ reflector telescope is adapted for public use and is often maintained by a local astronomy club. This program has facilitated new collaborations between amateurs and libraries and allowed new audiences to experience looking through a telescope.

Still other efforts engage amateurs in sharing their passion for astronomy in additional creative ways. The goal of Astronomers Without Borders (https://astronomerswithoutborders.org/) is to connect astronomy enthusiasts within communities and around the world. They organize and connect volunteers around significant astronomical events to facilitate the sharing of resources to empower enthusiasts and educators. Mike Simmons, the founder, is convinced that amateur astronomers represent the largest opportunity for outreach worldwide. He reports that Astronomers Without Borders has been able to reach tens of millions of people through the work of amateurs. As he described, "I've never found a country that doesn't have amateur astronomers." He shared that amateurs in developing countries lacking telescopes do outreach through posters and exhibits, driven by their excitement to share their knowledge about astronomy. One project connected to Astronomers without Borders, the World at Night (www.twanight.org), is active in astronomy at the international level and produces nightscape photos and videos of landmarks against celestial attractions as a bridge between art, science, and culture. The project produces and shares these images to inspire the public to learn and care more about the night sky, to highlight the importance of preserving dark skies and the natural night environment for all species, and to produce visual content for astronomy educators and other communicators.

One other unique outreach initiative is NASA's Solar System Ambassadors program (https://solarsystem.nasa.gov/solar-system-ambassadors/events/). Participants are space enthusiasts from all walks of life and interest who tell the story of NASA in a way that is meaningful to their local communities. Only a small fraction of ambassadors are observers. Born out of the legacy of NASA's Galileo Ambassador program to engage teachers in increasing public understanding of the Galileo mission to Jupiter, the current program has broadened in the past 20 years beyond working with just teachers to include individuals from across society, formal and informal educators, homeschool parents, those in STEM careers, and even individuals from ministries and ranging in age from 18 to 100. The program provides virtual training and resources for different NASA missions. Ambassadors are chosen via a yearly application, and the program looks for individuals who are well connected in their community. Ambassadors are in every US state, Puerto Rico, Guam, the US Virgin Islands, and overseas US military bases and consulates. Applications have increased coincidentally with major astronomy-related events, the release of a popular astronomy movie (for example, *The Martian* in 2015), the total solar eclipse visible across the United States in 2017, and the 50th anniversary of Apollo 11 in 2019 (Ferrari, 2020). To date, ambassadors have supported over 50,000 events throughout the United States.

## Amateurs as Researchers

There is a large and growing number of amateur astronomers who contribute to the scientific research community. Whereas amateurs have played such a role for more than a century, from around the 1990s, computer-controlled telescopes along with CCD cameras increased the ability of amateur researchers with finite budgets to contribute meaningfully to science (Baruch, 1992). It was through the help of dedicated deep sky astrophotographers that the stellar streams from galaxy mergers were detected, providing evidence in support of the galactic merger theories in 2008 (Martinez-Delgado et al., 2008). Robotic telescopes now allow urban astronomers to engage with remote dark skies and conduct more observation beyond their backyard or local area (Gomez & Fitzgerald, 2017). Examples of areas where amateur research is particularly active or emerging are eclipsing binaries (e.g., Guinan et al., 2012), exoplanets (e.g., Sarva et al., 2020; Zellem et al., 2020), and double stars (e.g., Freed et al., 2017). These opportunities leverage the fact that amateurs collectively can contribute enormous amounts of observing time and observing power stretches across the globe, and they have increasingly more advanced equipment. In this sense, amateur astronomers are a special class of citizen scientists, focused on using developed technical skills to contribute observations and analyze data to produce knowledge about the cosmos (for more on citizen science activity in space science, see Chapter 10 by Lucy Fortson in this volume).

Amateurs often do research through "pro-am" collaborations in which professional astronomers work directly with, or use data collected by, amateur astronomers, providing strong benefits for both amateurs and professionals (Henden, 2011; Price, 2012). Amateurs have supported planetary exploration including the monitoring of Jupiter in connection with the Juno mission (Orton, 2012) and other planetary science research (Mousis et al., 2014). Exoplanet Watch (https://exoplanets.nasa.gov/exoplanet-watch/about-exoplanet-watch/), managed by the Jet Propulsion Laboratory on behalf of NASA's Universe of Learning, engages amateur astronomers in observing transiting exoplanets with their own telescope or with archival data (Zellem et al., 2020). Observers reduce the data themselves and submit to a database managed by the American Association of Variable Star Observers (AAVSO). During the current beta-test phase, recruitment is done through word of mouth and the team speaking to clubs. The project goal is to continually monitor 200 high priority targets and to keep them up to date. To support users, volunteers have created tutorial videos, are translating material into languages beyond English, and provide support to other users in the community. Important to the project is that all data products are available to anyone and that proper attribution is given to the amateurs who provide the data and make the research possible.

Several organizations work to facilitate amateurs and pro-am collaborations. The AAVSO (https://www.aavso.org/) is arguably one of the most pervasive and important organizations for amateur researchers. The AAVSO enables connections to be made by providing projects, training, databases, and events to connect individuals across the world. Current membership comes from nearly 50 countries and is growing each year. The AAVSO curates a group of databases for all types of variable star photometry (including exoplanet transits) and spectroscopy and provides courses, workshops, and seminars as well as tools for data reduction and data analysis. In addition to the paid staff who make the organization run, AAVSO is run by a loyal and passionate army of volunteers. Often, based on the needs of a professional or pro-am collaboration that needs additional measurements of a target object to support observations or confirm a new discovery, the AAVSO puts out a call for a targeted observing campaign of different objects. Each observer reduces their own data and uploads it to the AAVSO database for anyone to access. Outcomes of this research are discussed at the AAVSO's meetings and are regularly presented in the professional scientific literature. The *Journal of the American Association of Variable Star Observers* (JAAVSO) is a peer-reviewed publication dedicated to variable star astronomy that is open to articles submitted by members of the AAVSO community and open to any and all amateur and professional members of the variable star research and observation community.

The Society for Astronomical Sciences (SAS) (http://www.socastrosci.org/), with over 100 members, is an organization made up mainly of advanced amateur astronomers who collect data on exoplanet transits, asteroid photometry, luminous star surveys, optical signals from X-ray transients close binaries, and many other astronomical phenomena. SAS provides a venue for the subset of amateur astronomers who are interested in doing research and contributing their work to the greater body of human knowledge. Many members are engineers who love to tinker and work on designing and prototyping and building new tools for imaging and spectroscopy. These amateurs get together at an annual symposium in Southern California to give talks, share their stories, learn from each other, and build and maintain their community. The research presented at their annual symposium is collected in the proceedings and freely available to the worldwide community through their website.

The Pro-Am Collaborative Astronomy (PACA) Project, run by astronomer Dr. Padma Yanamandra-Fisher from the Space Science Institute in Boulder, Colorado, grew from a successful NASA observational campaign to characterize Comet ISON in 2013. The initial campaign brought together amateur astronomers, educators, students, outreach coordinators, and astronomy journalists. Today, PACA has multiple observing campaigns that engage amateurs around the world collecting data on comets, planets, observing techniques (polarization and spectroscopy), and modeling. Observing campaigns are managed mainly on Facebook and other social media. New members are always welcome to join The PACA Project (https://www.facebook.com/ThePACAProject/).

In addition, the *Journal of Double Star Observations* (JDSO) (http://www.jdso. org/) is a peer-reviewed journal established in large part to support the amateur astronomy community participating in pro-am collaborations around the astrometric measurements of double stars. Most of the papers in the JDSO come from advanced amateur astronomers with the time, equipment, knowledge, and commitment to measuring double stars. The journal also publishes articles on instrumentation as the technology improves and new techniques and equipment are designed and applied to this sort of research.

Sometimes amateurs work through their own channels to contribute to the field. Recently, on September 5, 2020, an avid amateur astronomer and astrophotographer, Christopher Go from the Philippines, captured an image of a possible new storm on Jupiter's North Temperate Belt (NTB) and put out a call on social media for planetary astrophotographers around the world to try to capture more images as the nighttime side of Earth headed west. "ALERT! Possible new NTB plume. If you look at these two images, there is a bright spot setting on the NTB on the second image (12: 23UT). The first image (11:13UT) does not show anything significant. Please try to image this region around CM2: 277 to see if this is real! Is this an outbreak or just a high altitude cloud that is bright at the terminator? Those in Africa and Europe, please take methane band images of this region." Clearly, amateur astronomy is a global endeavor within a global community.

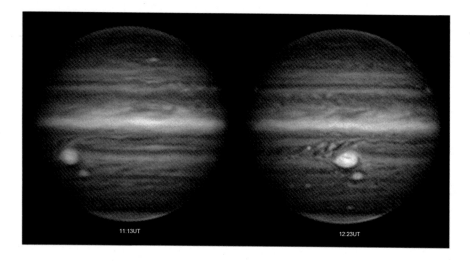

*Go Christopher*

Amateur astronomers also participate in the development of hardware and software for the field. Especially in the early days, much of the astronomical software that was developed for amateur use was developed by amateurs themselves. Sometimes this quickly evolved into formal businesses supporting amateurs. Some examples include TheSkyX or ACP for observatory control, MaximDL for image collection and processing, and tools like Peranso (Paunzen & Vanmunster, 2016), VStar (Benn, 2012), and C-Munipack for photometric processing and analysis. Others, like AstroImageJ (Collins et al., 2017) or EXOTIC (Zellem et al., 2020), were developed by amateurs and professionals for use in pro-am collaborations.

In addition to conducting their own research, amateurs sometimes aid students with their own research. Research experiences for undergraduate (Krim et al., 2019) and high school students (Fitzgerald et al., 2014, p. 31) are burgeoning, and much has been done in the past 5 years to bring astronomical research experiences to students and educators around the globe. Programs such as Our Solar Siblings (https://www.oursolarsiblings.com/) (Fitzgerald, 2018) in Australia, the Boyce Research Initiative and Education Foundation (http://boyce-astro.org/) (Boyce & Boyce, 2017), and the Institute for Student Astronomical Research (https://www.in4star.org/) in the United States have been teaching students and teachers how to conduct simple research projects in double star astrometry, exoplanet transit light curves, and variable star photometry. The major goals of these types of programs are to help students learn the process of science as well as how to communicate science through writing and public speaking. Critical to these initiatives is one-on-one mentoring of a knowledgeable adult, often an amateur astronomer, to guide the students from an initial through a full research project. Amateurs make this possible due to their numbers and their passion for sharing.

## Supportive Organizations and Resources

As already mentioned, amateurs are well supported in their pursuit of research. Many additional organizations and resources at the local, national, and international levels exist to nurture amateurs whether they wish to perform research, explore independently, or conduct public outreach.

### Astronomy Clubs, Societies, and Associations

Astronomy clubs, most often run by dedicated volunteers, provide structure, events, and a social network for individuals to engage in astronomy and serve specific cities, towns, or localities. Many amateur astronomers are members of formal clubs, although the exact percentage is unknown. Of the 850 respondents to a survey conducted by the Night Sky Network for amateurs engaged in outreach, more than 55% reported being a member in a formal astronomy club (White & Prosper, 2014). Some groups flitter in and out of existence quickly, whereas some have been going strong for many decades and a few are nearing a century. Clubs, along with local or regional astronomy societies and associations, are important to amateur astronomy as a nexus for like-minded individuals to gather, participate in activities together, learn and share knowledge, and engage in outreach and advocacy.

Current data from the Night Sky Network indicate that there are nearly 600 astronomy clubs in the United States dedicated to outreach. Indeed, the number of local clubs, societies, and associations engaged in public outreach is too great to discuss many in great detail here, but a few examples are in order. Members of the San Mateo County Astronomical Society (http://www.smcasastro.com/) participate in major science outreach programs throughout the San Francisco Bay Area. In addition to monthly club meetings that feature an invited speaker, they participate in several large events including the Stanford Linear Accelerator kids' night, an 8-hour event which serves 700–800 people each year. Volunteers bring their equipment to show views of the sun through solar scopes until it sets, followed by planets and other night sky objects until closing time. These events are filled with an electric energy as kids run around excitedly learning about all different manners of physics phenomena and then excitedly look at Jupiter and its moons or Saturn in different telescopes. They are just as invigorating for the amateur astronomers who revel in the excitement that the members of the community display about their shared passion. Meanwhile, on the East Coast of the United States, the Amateur Astronomers Association of New York (https://www.aaa.org/), offers public observing and a wide range of classes for members ranging from introductory astronomy to advanced astrophotography. Other activities include local lectures by professional astronomers, support for astrophotography, and a large event, StarFest, held twice a year in Central Park, that sees thousands of visitors.

Some amateur astronomy societies support the operation of local observatories. The Robert Ferguson Observatory (https://rfo.org/) in Sonoma County, California, has nearly 750 members and 8000–10,000 visitors per year. The observatory provides a dark sky site and an outlet for outreach so that amateurs can share their passion with the public. The observatory houses a 40″ telescope designed and built by volunteer docents purely for the public. There are about 120 active volunteers, who dedicate several hours each week to public outreach events. In addition to attending public events, visitors can rent a night at the observatory, complete with docents to give night sky tours and presentations. The observatory also supports a contest, "Striking Sparks," in which middle-school students write an essay about their interest in astronomy and the winner receives a telescope and participates in public outreach events. On the other side of the Pacific, the Mount Burnett Observatory (https://mbo.org.au/) is a nonprofit society outside of Melbourne, Australia. Of the 400 members, over a third are female and a quarter are children and teens. Part of this success in diversity, says President James Murray, is attributed to the young observers group, which gives younger members experience and opportunities to engage in outreach and other events. Additionally, the observatory makes strong connections in the surrounding community by bringing telescopes to the general public and organizing events to connect with schools, local universities, and the public.

## Umbrella Organizations

In addition to local groups are national and international organizations that support amateurs and clubs in their pursuits as independent explorers, outreach agents, and researchers. These umbrella organizations provide content, workshops, events, and networks to further the reach and connections to and for clubs. The Night Sky Network (NSN), for one, is a US-based coalition of amateur astronomy clubs supported in their outreach activities by NASA's Jet Propulsion Laboratory and the Astronomical Society of the Pacific (https://nightsky.jpl.nasa.gov/). The NSN supports clubs by providing a central place to advertise their events through an online searchable calendar as well as online training and resources to engage in outreach activities. Member clubs also receive support in managing their club, including membership, volunteer hours and dues, communication, and event management. Established clubs with at least 15 members and a history of outreach are eligible to apply for free membership.

The Astronomical League (astroleague.org) is another umbrella organization of amateur astronomers in the United States that has over 300 member clubs and about 18,000 individual members, most of whom are members in a local astronomy club. They publish a quarterly magazine for members, *The Reflector*, that shares updates about member clubs, beautiful pictures taken by members, information about awards given out, tips for observing, and science articles. In addition, members have access to over 75 guided observing programs that they can earn a certificate and pin for completing. The Astronomical League also provides a variety of awards for individuals who promote and advance astronomy and for young amateur astronomers. Each year they host a national convention, the ALCON (Astronomical League Convention), which includes lectures, events, and a banquet over 4 days. This convention brings together amateur astronomers to learn about the latest discoveries in astronomy, exchange ideas and techniques with other amateurs, and socialize with like-minded enthusiasts. The organization coordinates a worldwide Astronomy Day each fall and spring, which convenes clubs, museums, and observatories for coordinated observing events. The website provides resources for advertising and hosting events as well as awards for participating clubs.

The International Dark Sky Association (IDA) (https://www.darksky.org/) directly supports amateurs interested in advocacy work in their communities to improve night skies and overall quality of lighting on Earth. Members come from over 56 countries across the globe, and the impact of IDA's work comes from a large group of passionate volunteers who come from all walks of life. The paid staff of IDA support volunteers by providing information as well as resources to engage communities in discussion to reduce light pollution. There are many success stories of amateurs working with IDA that include changing city lighting policies and working with local, state, and national parks to become designated dark sky sites.

Amateurs and professionals mutually aid one another through the International Astronomical Union (IAU) (https://www.iau.org/), the largest international association of professional astronomers, with over 12,000 members across 107 different countries (https://iau.org/administration/about/). The IAU views the relationship between professionals and amateurs as paramount and actively works to strengthen the bond, not only for the important contributions of amateur astronomers to scientific research but also because amateur astronomers possess vast knowledge in astronomy and strong links with society at national and local levels. Activities to connect professional and amateur astronomers at the outreach level are led through the IAU Office for Astronomy Outreach (OAO), established in 2012, which focuses on engaging with the public, providing access to astronomical information, and widely communicating the science of astronomy (OAO, 2020). The OAO network of national outreach coordinators in over 135 countries is largely composed of amateur astronomers. Over the past decade, the IAU has proactively strengthened its relations with the worldwide amateur astronomical communities and in 2019 held the IAU 100th year anniversary celebration, with direct engagement of up to 10 million people in 143 countries (Rivero González et al., 2019). These actions owed their success largely to the engagement of amateur astronomical societies around the world. The celebrations highlighted the first IAU Amateur Astronomers Day celebrated on April 13, 2019, with the intention of furthering relationships between the IAU and amateurs and their organizations.

## Events, Online Forums, and Popular Publications

Observing and imaging often blends with outreach and research. Among observers and imagers are those who love to share what they do and how they do it, connecting with fellow observers through workshops, conferences, and star parties. For example, for 15 years, the Northeast AstroImaging Conference (NEAIC) has provided a venue where those who are interested in doing astrophotography can gather to learn and share their knowledge and experience. Inherent in their philosophy is making sure that anyone at any level will find learning and networking opportunities at the annual event in New York. It is one of the few conferences where the telescope, camera, and other equipment companies set up booths as exhibitors but are not allowed to sell their equipment; it is strictly a place to come and learn about techniques and what is available for the hobby.

Online platforms have also become central for amateur astronomers to communicate with each other and to share their discoveries. The Astro Imaging Channel (TAIC, www.theastroimagingchannel.org), a nonprofit organization, is a collaborative effort of avid amateur astrophotographers who meet weekly to discuss everything pertaining to astrophotography, from equipment to image processing to participating in scientific research. They invite professional astronomers for discussions and stream their talks live on YouTube, often having 50—70 participants viewing live, chatting, connecting with fellow astrophotographers, and asking questions. The channel has over 9000 subscribers; the videos are all available to watch for free and have collectively garnered almost 1 million views as of September 2020. Astrophotographers can share their images on sites such as AstroBin (https://welcome.astrobin.com/), the self-proclaimed "home of astrophotography," where others can go to learn about techniques for advancing their own craft. Participants share the details associated with the images they post, such as equipment, exposure times, and other camera settings. In addition, online astronomy forums, such as Cloudy Nights (www.cloudynights.com), allow the community to ask and answer questions relating to anything within the amateur astronomy world, from how to set up telescopes, to the newest technologies for image capture and processing, to organizing meetups at conferences. While many people post and participate in discussions on the forum, there are untold multitudes who simply read and learn from the posts and discussions.

Print magazines also remain an integral source of information and support for the amateur community. In a time of decline in magazine subscriptions, *Sky & Telescope* (https://skyandtelescope.org/) has nearly 50,000, and *Astronomy* (https://astronomy.com/) has nearly 75,000 subscribers to their print magazines, respectively, with even larger numbers of digital subscriptions and web visitors. Both publications support observing events, work with clubs, and are sources for relevant, trusted, and high-quality information for a broad range of space enthusiasts. *Sky & Telescope* remains a "go to" resource for amateur astronomers and space enthusiasts with a serious interest in astronomy. The content supports avid knowledge seekers and includes accessible technical pieces about what is visible in the sky and what is happening in the science of astronomy. *Astronomy*'s goal is to provide a variety of ways to interact with the enormous spectrum of individuals interested in astronomy. Both magazines are increasing online content to connect to a broader readership and younger audiences.

## Challenges and Innovations due to the COVID-19 Pandemic

Early 2020 saw worldwide shutdowns due to the spreading of COVID-19 and the health crisis that ensued. As businesses and schools were shut down across the globe, amateur activities were hastily canceled. Many clubs realized that holding any in-person events would put members at risk but that online programs might pose technological barriers to older members. As soon as it became clear that the shutdowns and shelter-in-place orders around the globe were going to last more than a few weeks, many individuals and organizations began thinking about how to provide their astronomy and outreach services virtually.

Although some amateur astronomers chose to continue observing on their own, whole new online communities of amateurs developed. With extended shutdowns, organizations continued their engagement through online lectures, workshops, and virtual star parties. While virtual star parties were introduced in 2013–14 by Universe Today through Google Hangouts, the concept of online star parties as a mainstream way for astronomy clubs to connect with their members is new to most organizations. Numerous astronomy clubs have begun having live viewing sessions through Zoom and other video conference programs, where amateur astronomers, from their own backyards, can connect their personal telescopes and cameras to their computer and do live imaging to share with each other and the public. There are technical challenges to this process, but clubs are working through these. Several large virtual star parties have been successful. The Explore Alliance Global Live Virtual Star Parties (https://explorescientificusa.com/products/explore-alliance-global-star-party), started by Scott Roberts at the start of the pandemic to supplement other outreach initiatives that he leads, has reported tens of thousands of views weekly while a virtual star party by Mount Burnett Observatory in September 2020 garnered 130,000 visitors via Facebook Live and over 10,000 comments with over 500 shares.

Similar to professional astronomy conferences, amateur societies are reimagining their conferences and workshops. Virtual events are reporting overall increased attendance, from club lectures to workshops and conferences. For example, the SAS held its annual symposium online as a joint meeting with the AAVSO over the course of five Saturdays in Spring of 2020 instead of the originally scheduled 3-day in-person symposium. Attendance almost doubled from the usual 120 participants to a little over 200. Meanwhile, AAVSO is conducting virtual events including its annual meeting and workshops held virtually over three weekends in November 2020. Whereas in-person events previously may have had 100 registrants, now online programs have up to 500 registrants from all over the world.

In addition to online innovation, some clubs have been experimenting with socially distant events. The Robert Ferguson Observatory has been successfully experimenting with socially distanced live star parties in which small family groups sit together under the dark skies, distant from other groups, and numerous docents provide tours of the night sky. They have recently set up an outdoor large television that can be connected to their robotic telescope so that they can do outdoor live imaging for socially distanced groups in the parking lot. Other amateurs have engaged members in their neighborhood with BYOB (bring your own binoculars) star parties. Amateur astronomers are also sharing their ideas online through networks such as the Night Sky Network and other online forums.

One area that has benefited from the coronavirus global shutdown has been that of telescope equipment sales. Almost everyone we talked to discussed the reemergence of night sky viewing as a hobby that has emerged during the shutdown. People have decided that while they are stuck at home, taking up the hobby of astronomy or upgrading their equipment or getting it out of storage is a good way to spend the time. Companies including Software Bisque, Stellarvue, Astro-Physics, and Cloudbreak Optics have all seen significant increases in orders for new equipment as well as calls for technical support on equipment that was purchased years ago.

The enduring questions are which of these innovations are here to stay and which will be only temporary. There have been many benefits for astronomers worldwide to connect online and reach new audiences. Amateurs have been able to attend events that they might not have been able to attend due to time of travel and expensive travel costs. Additionally, online lectures are seeing increased attendance by younger viewers who now can attend without disrupting the rest of their day to attend a lecture. Online events have decreased many barriers, such as families being able to attend late night events, while some younger family members go to sleep early. There are some concerns that the lack of in-person meetings will decrease the chance conversations that often lead to valuable collaborations. No matter what, amateurs will never give up looking up at the sky. The hope is that the best of online events can be leveraged to network amateurs across the world and support local connections more effectively.

## The Future of Amateur Astronomy

As a field, we need to attend to the documented barriers to increase diversity and inclusion for the future of amateur astronomy as a hobby and to ensure that members from across society are represented. A recent study, as well as our own interviews, investigated barriers to participation in amateur astronomy by ethnic and racial minorities and women. Reported barriers included concerns about personal safety in observing in the dark of night, postponed hobby development due to life circumstances, reduced social contact (such as feeling isolated as the only racial or ethnic minority or one of few women, leading to a lack of interaction or observing on one's own), lack of access (for example, light pollution in urban areas and lack of dark sky sites), and the prohibitive costs of buying a telescope to engage in the hobby. When asked about recommendations to encourage more individuals in their ethnic and racial communities to participate in astronomy, those interviewed shared the need for "(1) engaging young students, (2) using children's natural curiosity to spark interest in astronomy, (3) removing misconceptions or barriers regarding minority participation in astronomy, and (4) creating hobby spaces that are comfortable and welcoming to minorities." Respondents suggested the intentional use and celebration of role models exemplifying minorities engaged in STEM disciplines. Regarding the participation of women, the authors of the published study stressed the importance of attending to concerns for personal safety (Corin et al., 2018). Although more research is needed around increasing diversity in amateur astronomy, similar practices to increasing the diversity of STEM professionally (Reyes, 2012; Tsui, 2007) may be useful in thinking about inclusion of different individuals in amateur astronomy. As an example, the AAVSO is working to address the overall lack of diversity in their membership through several initiatives including a diversity committee and ambassador program composed of young energetic individuals committed to outreach and inclusion.

Furthermore, there are tensions between amateur and professional astronomy to address. While professional astronomers engage in outreach, and at a greater rate than in other professional sciences (Dang & Russo, 2015), there are simply not enough of them to undertake outreach at a large scale. Additionally, amateurs providing outreach further demonstrate broadly to audiences that astronomy can be conducted by individuals from all backgrounds and walks of life (Slater, 2007). Amateurs fill an important niche in outreach due to being greater in number and often having a close connection to their community. It is important for professionals to recognize that amateurs are strong mediators to support the world of professional science by sharing information and making more personal connections with many more humans on Earth (Storksdieck & Berendsen, 2007; Simmons, 2020; Wenger, 2011).

Despite concerns of professional astronomers about the quality of data collection collected by amateurs, they have demonstrated that as a group they provide valid, reliable, and consistent observations that are of high value to the astronomy community. There is also an ongoing concern from amateurs about contributing to the field of astronomy in an ever changing environment. As SAS president Bob Buchheim described, "The fear going around for 20 years is that the march of professional astronomy will make us obsolete." However, he describes that in the era of huge telescopes and enormous volumes of data coming in from ground- and space-based surveys, amateur researchers will be kept busy for a long time, as more targets of interest are being identified without resources to follow up on them, and professional surveys cannot follow objects that change in hours or minutes. Overall, there will continue to be a need for and growth of amateurs as researchers.

Amateur astronomy provides the capacity to make enduring social and scientific connections between professionals, amateurs, formal education, and the public. It is a collection of small (and not-so-small) interdependent communities of practice (Lave & Wenger, 1991; Wenger, 2010) that add up to a social web of individuals who have a shared love of astronomy. Amateur astronomy has multiple entry points from childhood to retirement and a constellation of ways that they engage and support others. It is a vibrant hobby that is successfully attracting a new group of younger members through targeted engagement (see highlighted profiles as examples). Important to the growth and cultivation of amateur astronomy are individuals and organizations, composed of both amateurs and professionals, who actively provide opportunities for amateurs to grow in knowledge, experience, skills, and opportunities to do more. These individuals sit at a nexus between multiple groups and connect individuals and groups for an overall larger impact. Overall, amateurs continue to be important to the field of astronomy as independent explorers, outreach agents, and researchers.

## Postscript: Profiles

As younger amateur astronomers are key to the future of amateur astronomy, it is important to highlight some of the activities, outreach, and research that they are engaging in. Below we provide four profiles of highly engaged and motivated young amateur astronomers. Each of these individuals exemplify crossing the boundaries between independent explorers, outreach agents, and researchers and can serve as role models for a whole new generation of people interested in the hobby.

Molly Wakeling is currently a PhD student in physics at the University of California, Berkeley and has been interested in astronomy from a young age. When she was young, she fed her interest with books and TV shows. Five years ago, she was given a telescope which she used to make observations of the night sky. After seeing Saturn, she was hooked! She joined the local astronomy club in which older members gave her gear and supported her in astrophotography. She gets great satisfaction from her own astrophotography work. In addition, she has been active in outreach, hosting star parties and other events for the public and scouts. She also writes blog posts about astronomical objects to help people increase in their appreciation of the universe. She currently owns five telescopes and a pair of binoculars. She reports that she often gets a lot of attention from the public as a young woman who is knowledgeable about astronomy. She is an active member of the AAVSO and Exoplanet Watch and is an Explore Alliance Ambassador. She is passionate about diversifying the field and serves as an AAVSO ambassador. Her many roles as an amateur astronomer allow her to bring her passion and her knowledge to contribute to both outreach and research in substantial ways.

Molly Wakeling with two of her telescope rigs and her camper at a dark sky site outside Sacramento, CA. *Molly Wakeling.*

Captured with a Sony a7s astro-modified camera with a Rokinon 135 mm f/2 lens at f/2.8 on a Sky-Watcher Star Adventurer platform with 30 min total exposure. Taken in Chile's Atacama Desert. *Rho Ophiuchi Complex. Molly Wakeling.*

Lauren Herrington is a 19-year-old observer and spectrographer whose interest in astronomy first developed when she borrowed a toy telescope from a family friend. Her mother encouraged her to attend the local astronomy club's meetings and star parties, facilitating her learning and nursing her growing passion for outreach. Unfortunately, due to her age, it would be several years before a community of astronomers would notice and accept her. Despite the lack of external support, Lauren persevered, observing and teaching about the night sky on her own. She has been in love with spectra ever since reading a book about the Sloan Digital Sky Survey and spent years playing with catalog data and building her own spectroscopes. In 2019, she attended Sacramento Mountains Spectrography Workshop and has since developed a data collection method, which allows beginners to record spectra without expensive equipment. She now works with the AAVSO in presenting community workshops and is writing a paper to describe her drift scanning method in detail. Lauren strives to make meaningful contributions to the study of astronomy and enjoys using outreach to inspire others, making a special point to be available to beginners—especially those who are isolated as a result of their age or gender.

Lauren Herrington pauses to find the planets while setting up for a star party. *Lisa Herrington Morgan.*

Dark sky site self-portrait. *Lauren Herrington.*

Ryan Caputo is a young avid amateur astronomer who first became involved in astronomy at the age of 16 when he took a course at the Center for Talented Youth at Johns Hopkins University. He was so inspired by the course that he pulled out his 2½" reflecting telescope that had never been seriously used and started peering at objects in the night sky. He received a 10″ Dobsonian telescope for his birthday and then "spent every clear night outside with it." He spent considerable time observing and finding fainter objects. As a sophomore in high school, he took an astronomy research course and started publishing his work. In 2020, he worked with the Fairborn Institute Robotic Observatory (FIRO) to help develop a training manual on how to do speckle interferometry on close double stars with the FIRO telescope, publishing several papers in the process. This Bubble Nebula image was taken the night before he left for college and posted on his AstroBin site. In the description on the page, Ryan says: "This is the last photo from me for a while. I am headed off to college; tomorrow, actually. I won't be taking my telescope with me. I find this picture a most fitting end to a wonderful hobby. We will see where life takes me. The joys, the pain, the late-night computer crashes and telescope misbehavings, all summed up nicely in this one photo. All the techniques I've learned, applied. All the effort, shown in this one photo."

Bubble Nebula taken with a total of 12 hour of exposure. *Ryan Caputo.*

Ryan Caputo in his backyard with his telescope, a 6-inch classical Cassegrain and ZWO ASI 1600 mm camera riding on an iOptron CEM60 mount. This system specializes in deep sky astrophotography. *Ryan Caputo.*

Indra Firdaus is the current chair of the Jakarta Amateur Astronomy Club (HAAJ). He has loved the stars since he was a child and joined the club when he was 16 years old. He, like most others in Indonesia, does not own a telescope, and joining the HAAJ gave him the opportunity to use a telescope for the first time. In high school, he sketched images of Venus and participated in the International Year of Astronomy. He has been a member of Astronomers without Borders since 2011. He is working with others across Indonesia to develop a national network for the future of astronomy development. He is very active in outreach and participates in numerous outreach events per year. The club holds biweekly meetings with professional and amateur astronomy speakers and hosts star parties and expeditions for special events. Recently the HAAJ has taken to holding its meetings virtually.

Indras at the Borobudur temple, Magelang, Central Java. *Indra Firdaus.*

Photos of the total solar eclipse taken on March 9, 2016, in Kabonga, Banawa District, Donggala Regency, Central Sulawesi Province, Indonesia, using a digital single-lens reflex camera and standard lens. The left photo is a single image, while the right photo is a montage of all the eclipse phases. *Indra Firdaus/HAAJ.*

## Acknowledgments

We are grateful to the many people who agreed to share their stories in constructing this chapter. They include Joshua Alfred Anderson, Mutoha Arkanuddin, John Barentine, Kelly Beatty, Brian Berg, Stephen Bisque, Pat Boyce, Bob Buchheim, Lina Canas, Ryan Caputo, Dave Eicher, Kay Ferrari, Don Ficken, Indra Firdaus, Andy Fraknoi, Eko Hadi Gunawan, Tolga Gumusayak, Ruskin Hartley, Lauren Herrington, Stella Kafka, Dave Kensiski, Molly N. Kgobathe, George Loyer, Naoufel Ben Maaouia, Terry Mann, Vic Maris, Zied Mejri, Yasmine M'hirsi, Gaone Edwin Mogaetsho, Nonofo Mogopodi, Bob Moore, Amjed Mouelhi, James Murray, Martin Ratcliffe, Muhammad Rayhan, Robert Reeves, Scott Roberts, Mike Simmons, Yassine Tahri, William Christian Tchaptchet, Babak Tefreshi, Muhammad Thoyib, Moletlanyi Tshipa, Bill Tomlinson, Molly Wakeling, Vivian White, Padma A. Yanamandra-Fisher, Rob Zellem, and five individuals who wish to remain anonymous.

## References

Baruch, J. E. (1992). Robots in astronomy. *Vistas in Astronomy, 35*, 399–438.

Benn, D. (2012). Algorithms+ observations = VStar. *Journal of the American Association of Variable Star Observers, 1*, 10.

Boyce, P., & Boyce, G. (June 2017). A community-centered astronomy research program. In *Society for astronomical sciences annual symposium* (Vol. 36, pp. 107–122).

Cameron, G. L. (2010). *Public skies: Telescopes and the popularization of astronomy in the twentieth century. Doctoral Dissertation.* Iowa State University. https://lib.dr.iastate.edu/etd/11795.

Collins, K. A., Kielkopf, J. F., Stassun, K. G., & Hessman, F. V. (2017). AstroImageJ: Image processing and photometric extraction for ultra-precise astronomical light curves. *The Astronomical Journal, 153*(2), 77.

Corin, E. N., Jones, M. G., Andre, T., & Childers, G. M. (2018). Characteristics of lifelong science learners: An investigation of STEM hobbyists. *International Journal of Science Education, Part B, 8*(1), 53–75.

Dang, L., & Russo, P. (2015). *How astronomers view education and public outreach: An exploratory study. arXiv preprint arXiv:1507.08552.*

Ferrari, K. (2020). *Personal communication.*

Fitzgerald, M. T., McKinnon, D. H., Danaia, L., Cutts, R., Salimpour, S., & Sacchi, M. (2018). Our solar siblings. A high school focused robotic telescope-based astronomy education project. *RTSRE Proceedings, 1*(1), 221–235. https://ro.ecu.edu.au/ecuworkspost2013/5236/.

Fitzgerald, M. T., Hollow, R., Rebull, L. M., Danaia, L., & McKinnon, D. H. (2014). *A review of high school level astronomy student research projects over the last two decades.* Publications of the Astronomical Society of Australia.

Freed, R., Fitzgerald, M., Genet, R., & Davidson, B. (June 2017). an overview of ten years of student research and JDSO publications. In *Society for astronomical sciences annual symposium* (Vol. 36, pp. 131–136).

Gada, A., Stern, A. H., & Williams, T. R. (2000). What motivates amateur astronomers?. In *Amateur-professional Partnerships in astronomy* (Vol. 220, p. 14).

Gomez, E. L., & Fitzgerald, M. T. (2017). Robotic telescopes in education. *Astronomical Review, 13*(1), 28–68.

Guinan, E. F., Engle, S. G., & Devinney, E. J. (2012). Eclipsing binaries in the 21st century—opportunities for amateur astronomers. *Journal of America Association of Variable Star Observers, 40*, 467–480.

Henden, A. A. (2011). Amateur community and "citizen science. *Proceedings of the International Astronomical Union, 7*(S285), 255–260.

Kannappan, S. (2001). *Border trading: The amateur-professional partnership in variable star astronomy*. Cambridge, MA: Harvard University.

Krim, J. S., Coté, L. E., Schwartz, R. S., Stone, E. M., Cleeves, J. J., Barry, K. J., Burgess, W., Buxner, S. R., Gerton, J. M., Horvath, L., & Keller, J. M. (2019). Models and impacts of science research experiences: A review of the literature of CUREs, UREs, and TREs. *CBE-Life Sciences Education, 18*(4), ar65.

Lave, J., & Wenger, E. (1991). *Situated learning: Legitimate peripheral participation*. Cambridge University Press.

Martínez-Delgado, D., Penarrubia, J., Gabany, R. J., Trujillo, I., Majewski, S. R., & Pohlen, M. (2008). The ghost of a dwarf galaxy: Fossils of the hierarchical formation of the nearby spiral galaxy NGC 5907. *The Astrophysical Journal, 689*(1), 184.

Mousis, O., Hueso, R., Beaulieu, J. P., Bouley, S., Carry, B., Colas, F., Klotz, A., Pellier, C., Petit, J. M., Rousselot, P., Ali-Dib, M., Beisker, W., Birlan, M., Buil, C., Delsanti, A., Frappa, E., Hammel, H. B., Levasseur-Regourd, A. C., … Widemann, T. (2014). Instrumental methods for professional and amateur collaborations in planetary astronomy. *Experimental Astronomy, 38*(1–2), 91–191.

*Office for astronomy outreach*.(2020). https://www.iau.org/public/. (Accessed 15 September 2020).

Orton, G. (2012). A new mission-supporting era of amateur astronomy: The Juno mission and the role of amateur astronomers. *European Planetary Science Congress*. EPSC2012-288.

Paunzen, E., & Vanmunster, T. (2016). Peranso—Light curve and period analysis software. *Astronomische Nachrichten, 337*(3), 239–245.

Percy, J. R. (1998). The role of amateur astronomers in astronomy education. In *International astronomical union Colloquium* (Vol. 162, pp. 205–210). Cambridge University Press.

Plummer, J. D., & Small, K. J. (2013). Informal science educators' pedagogical choices and goals for learners: The case of planetarium professionals. *Astronomy Education Review, 12*(1), 1–16.

Pompea, S. M., & Russo, P. (2020). Astronomers engaging with the education ecosystem: A best-evidence synthesis. *Annual Review of Astronomy and Astrophysics, 58*.

Price, M. W. (2012). *Stellar connections: Explorations in cultural astronomy—Pt. 2. Transcription of lecture*. https://wiki2.org/en/White_Earth_Tribal_and_Community_College. https://www.youtube.com/watch. Lecture on video.

Reyes, Marie-Elena (2012). Increasing diversity in STEM by attracting community college women of color. In B. Bogue, & E. Cady (Eds.), *Apply research to practice (ARP) resources*. http://www.engr.psu.edu/AWE/ARPResources.aspx.

Rivero González, J., Dishoek, E., Russo, P., Canas, L., & Downer, B. (2019). *IAU 100th anniversary celebrations: Final report*. https://www.iau.org/static/archives/announcements/pdf/iau100-final-report-ann20019.pdf. (Accessed 15 September 2020).

Salimpour, S., Bartlett, S., Fitzgerald, M. T., McKinnon, D. H., Cutts, K. R., James, C. R., Miller, S., Danaia, L., Hollow, R. P., Cabezon, S., Faye, M., Tomita, A., Max, C., de

Korte, M., Baudouin, C., Birkenbauma, D., Kallery, M., Anjos, S., Wu, Q., … Ortiz-Gil, A. (2020). The gateway science: A review of astronomy in the OECD school curricula, including China and South Africa. *Research in Science Education*, 1–22.

Sarva, J., Freed, R., Fitzgerald, M. T., & Salimpour, S. (2020). An exoplanet transit observing method using LCO telescopes, exoRequest and astrosource. *Astronomy Theory, Observations and Methods, 1*(1).

Simmons, M. (2020). *Personal communication*.

Skelton, R. (2013). Designing minimal-mass tensegrity telescopes of optimal complexity. *Astronomy, SPIE Newsroom*. https://doi.org/10.1117/2.1201302.004699

Slater, T. F. (2007). Working with schools: Teaching student on their own turf. In M. Gibbs, M. Berendsen, & M. Storksdieck (Eds.), *Science educators under the stars: Amateur astronomers engaged in education and public outreach* (pp. 18–30). Astronomical Society of the Pacific.

Slater, T. F., & Tatge, C. B. (2017). *Research on teaching astronomy in the planetarium*. Cham: Springer.

Storksdieck, M., & Berendsen, M. (2007). Attributes and practices of amateur astronomers who engage in education and public outreach. In M. Gibbs, M. Berendsen, & M. Storksdieck (Eds.), *Science educators under the stars: Amateur astronomers engaged in education and public outreach* (pp. 31–42). Astronomical Society of the Pacific.

Tsui, L. (2007). Effective strategies to increase diversity in STEM fields: A review of the research literature. *The Journal of Negro Education*, 555–581.

Waller, W. H., & Slater, T. F. (2011). Improving introductory astronomy education in American colleges and universities: A review of recent progress. *Journal of Geoscience Education, 59*(4), 176–183.

Wenger, E. (2010). Communities of practice and social learning systems: The career of a concept. In *Social learning systems and communities of practice* (pp. 179–198). London: Springer.

Wenger, M. C. (2011). *Free-choice family learning experiences at informal astronomy observing events*. Doctoral Dissertation, University of Arizona. http://hdl.handle.net/10150/202938.

White, V., & Prosper, D. (2014). *Internal report*. Night Sky Network.

Williams, T. R. (2000). *Getting organized: A history of amateur astronomy in the United States. Doctoral Dissertation*. Rice University. https://hdl.handle.net/1911/19569.

Zellem, R. T., Pearson, K. A., Blaser, E., Fowler, M., Ciardi, D. R., Biferno, A., Massey, B., Marchis, F., Baer, R., Ball, C., Chasin, M., Conley, M., Dixon, S., Fletcher, E., Hernandez, S., Nair, S., Perian, Q., Sienkiewicz, F., Tock, K., … Malvache, A. (2020). Utilizing small telescopes operated by citizen scientists for transiting exoplanet follow-up. *Publications of the Astronomical Society of the Pacific, 132*(1011), 054401.

# The Engagement Activities of ESTCube-1: How Estonia Built and Fell in Love With a Tiny Satellite

Arko Olesk, Mart Noorma

When the European Space Agency's Vega launcher took off from the Guiana Space Center in Kourou on May 7, 2013, it was early morning hours in Estonia. Despite this, a few dozen people had gathered in the main hall of the Baltic country's Tartu Observatory to follow the liftoff, filling the room with cheering and clapping once the rocket made its way into the dark night sky of French Guiana. They had been waiting for this moment for several days: the liftoff had been delayed day after day due to unfavorable upper-level winds. You could even say many in the room had been waiting for this moment for 5 years, since the inception of the ESTCube-1 student satellite program. Estonia's first CubeSat was now sitting on-board the Vega launcher, waiting to be released on orbit.

The room was full of tension, waiting to hear back from the satellite. A few hours after the launch, an email arrived from a Russian radio amateur with an attached recording of a fragmental signal. The team considered it likely that it originated from ESTCube-1, but for a full confirmation they wanted to hear from the satellite directly. At 10:35 a.m. local time, Tartu received a clear and strong signal. Another round of cheering went around the room when communications specialist Tõnis Eenmäe declared: "We have it. It's ESTCube-1." The tiny Baltic nation of 1.3 million people had become the 41st country in the world to send a satellite into Earth's orbit (Fig. 9.1).

We are able to reconstruct the events of that morning so precisely partly because they were followed and captured by the media: there were news cameras recording the emotions, reporters doing interviews, and a documentary crew following the satellite team over a longer period. The successful launch of the satellite became the main feature that night in the Estonian Public Broadcasting channel's 9'o clock news program, the most watched and most respected news show of the country.

The media attention to the satellite was not limited to its successful launch. The project enjoyed considerable public visibility throughout the duration of the project (2008–15) and also received several high recognitions such as Estonian Person of the Year 2013 (awarded by national newspaper *Postimees* to the project supervisor, and coauthor of this chapter, Mart Noorma), and Achievement of the Year (awarded by Estonian Public Broadcasting's news website in 2013 and selected by public vote).

Space Science and Public Engagement. https://doi.org/10.1016/B978-0-12-817390-9.00005-1

**FIG. 9.1**

Artist's rendering of ESTCube-1 in space. *SA Eesti Tudengisatelliit.*

The satellite became a cultural icon, its name and image being used in contexts as varied as the Estonian president's Christmas card and museum exhibitions.

In Estonia, the ESTCube project is often argued to have been a model example of good science communication: its visibility in the media allowed it to reach a large audience, the involved scientists were good at explaining the science, and it provided the public with positive stories about science. It is also a model of problem-based learning: the satellite was fully developed and built by students. While these characteristics are not unique among space projects globally, the ESTCube project had some features that make it a valuable example for the wider public engagement community.

This chapter describes the various engagement activities of the ESTCube-1 project as a model case of how a small research team is able to make a significant societal impact by dedicated engagement and communication work. We present how the ESTCube team designed and managed engagement of students and public communication activities, especially media relations, outlining their best practices. We also use the example of the satellite team to show how scientists can become skillful communicators.

We base this description on the combination of both insider participant and outside close observer perspectives, which included eight semistructured face-to-face interviews with the project members. One of us (Noorma) was in 2008 an associate professor in the University of Tartu who initiated and supervised the ESTCube-1 project and was one of the main actors in many communication activities. The other of us (Olesk) had contacts with the project as a science communication scholar (making the interviews and analyzing ESTCube's media coverage for his PhD thesis about changes in media-related attitudes and practices of scientists), as a journalist (writing science news and features for the Estonian daily *Postimees*), and as a science communication trainer (conducting media training courses for researchers including ESTCube team members).

## Overview of the ESTCube-1 Project

The idea of ESTCube was born as part of the space technologies study course at the University of Tartu. The course reader (Noorma) had become convinced that modern education needs a more problem-based approach and took inspiration from several other student satellite programs from around Europe.

The project was initiated in 2008 and announced publicly shortly thereafter. Although Estonia has a long history in astronomical research and was involved in developing some instruments for the Soviet space program, it had never designed or built a satellite. For this reason, there was initially some skepticism and concern in the scientific community that the task would be too complex for students. As a response, the team emphasized that the main focus of the project would be on the learning process.

However, the team still wanted the satellite to have a strong scientific mission as well. Looking for a potential mission, the team encountered the concept of the electric solar sail (E-sail), a novel propulsion mechanism proposed by the Finnish scientist Pekka Janhunen (2004). It was ambitiously decided that the Estonian satellite would become the first mission to test the sail in space; more specifically, the team wanted to test the deployment mechanism of an E-sail tether (Envall et al., 2014). The name of the satellite—ESTCube—has a dual meaning: it is both the Electric Sail Test Cube and the ESTonian CubeSat.

ESTCube-1 was a 1-unit CubeSat built according to the CubeSat standard, measuring $10 \times 10 \times 10$ cm and with a mass of 1.050 kg (Lätt et al., 2014, see Fig. 9.2). All subsystems and payloads were custom-built mostly using commercial off-the-shelf components. The software for the mission was developed by the students in collaboration with the software company CGI.

The satellite was launched in May 2013 and was operational for 2 years. The main scientific mission objective of the satellite was to perform the first in-orbit solar wind sail experiment, testing a novel technology expected to revolutionize space travel (Envall et al., 2014). A secondary technical objective of the ESTCube-1 mission was to take images of Estonia.

The satellite reached its full functionality by March 2014, but preparations for the scientific experiment revealed a strong magnetic disturbance, which did not allow spinning up of the satellite around the axis predetermined for the E-sail experiment ("ESTCube-1," n.d.). As the disturbance could not be reduced, the team tried to carry out the experiment with the satellite spinning around a different axis. When the experiment began on September 16, 2014, the downloaded images did not confirm

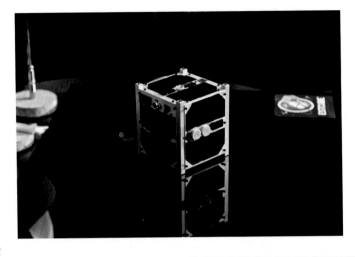

**FIG. 9.2**

ESTCube-1 satellite before being delivered to the launch provider. *SA Eesti Tudengisatelliit.*

deployment of the tether, the component of the E-sail to be tested with the mission ("ESTCube-1," n.d.). After several attempts, the team decided to spin up the satellite even further in hope of loosening any stuck components. However, even though the satellite achieved the fastest spin rate known to be deliberately reached by any spacecraft, the tether did not unreel. Further investigation and tests led to the conclusion that this was likely due to a failure of the tether reel motor ("ESTCube-1," n.d.).

The satellite did achieve its secondary objective: despite the magnetic disturbance and the resulting attitude control problems, the first full images of Estonia were taken in April 2014. In total, ESTCube-1 captured 300 photographs from space. After 2 years on orbit, the solar panels of the satellite no longer produced enough energy, and the last connection with the satellite was on May 19, 2015 ("ESTCube-1," n.d.).

Even though the satellite was not able to complete its main scientific objective, the satellite team considers the ESTCube-1 project a remarkable success, mostly because they achieved a number of other, nonscientific objectives for the project. The most important of these—and the main reason for the initiation of the project—was to increase the quality of education (Slavinskis, Reinkubjas, et al., 2015). During the course of the project, outreach also became increasingly important. In the beginning, outreach was mainly planned as a tool to attract students to the project. Later, after sensing great public interest for the project, the team expanded its communications activities to include more public engagement and media interactions. This was done because they perceived value in the mission to increase the visibility of and support for science and also to enable students participating in the project to acquire communication skills.

The next sections will discuss in detail the engagement activities of the ESTCube-1 team.

# Engagement With Students

The decision to initiate the project primarily for educational purposes defined the project's core principle: to develop all subsystems in-house and from the ground up despite not having any prior in-house experience with building satellites. This follows the precept of the problem-based learning approach as an effective learning tool: students work on a complex problem in collaborative groups to identify what they need to learn to solve it; they engage in self-directed learning and then apply their new knowledge to the problem (Hmelo-Silver, 2004). The skills gained through problem-based learning can be considered more in accordance with the demands of the modern job market than knowledge gained by traditional lectures. This concerns not only the specific skills related to satellite design and building but also transferable skills like project management, teamwork, and time management. Crucially, this also includes skills related to engagement such as managing public relations, interacting with journalists, and giving public presentations. For example, it was project policy that presentations at scientific conferences were to be made by students.

Another key principle the ESTCube team defined and followed for the project was that students would take the responsibility and lead their own work (Slavinskis, Reinkubjas, et al., 2015). Senior researchers (including Noorma, the project supervisor) were available as technical or academic advisors, but all subsystems and the project itself were led by students. Altogether, the number of students involved in the project at any stage was well over 200 (coming from 10 countries), and the main scientific publication describing the satellite had 89 student authors (Lätt et al., 2014).

The building of the team took place in several steps: the idea for the satellite was first devised by a group that included a handful of students. The first public presentations of the project (which included media coverage) attracted additional people, allowing the creation of a core team of about 10 students. This team started to search for a scientific mission for the satellite and, after selecting the E-sail experiment, started to develop all necessary subsystems and to recruit further students to the team (Fig. 9.3).

Students were recruited to the project via curricular activities such as a course about space technologies and thesis seminars. At the same time, a number of students also joined on a voluntary basis, leading to the group being heterogeneous in terms of level of education (high school, bachelor, master, and PhD students were all included) and the disciplines represented (physics, information technology, material sciences, engineering, etc.). The project also attracted several students from other universities.

At the same time, student turnover in the project was high: at least half of the students dropped out quickly after joining the project for various reasons. Finding good and dedicated student leaders for all subsystems was an especially long process, taking over 2 years. It quickly became clear that motivating students was crucial (Slavinskis, Pajusalu, et al., 2015). The core management team employed various strategies to keep the students engaged with the project. One was giving students a high level of responsibility and involving them in decision-making to

**FIG. 9.3**

ESTCube-1 team photo. Mart Noorma is in the first row, fourth from right. *SA Eesti Tudengisatelliit.*

give them a sense of ownership of the project. Additionally, choosing a challenging scientific mission (E-sail) over a simple one helped to create the sense of actual contribution to science and the future of space travel. Motivation strategies also used arguments emphasizing individual benefits such as the possibilities of a scholarship or an opportunity to use the work on the project in one's final thesis. Finally, the management also used media interest to argue to the project members that their work on the project was appreciated by society. The active facilitation of media interactions led to another motivating outcome, as explained by one team member:

> *The decision we made in the beginning to speak about the project publicly and interact with media certainly put additional responsibility to the team and created some stress. But on the other hand it also made us take the responsibility to really finish the project and not to give up. (Systems engineer)*[1]

Another key principle for the project was collaboration with industry and with international research institutions. These collaborations were mostly student-led and also functioned as a motivator, offering learning opportunities beyond the official university curriculum and thereby increasing the likelihood of employment for

---

[1] The quotes here and in the following sections come from eight in-depth interviews made by one of the authors (Olesk) with members of the ESTCube management team and leaders of several subsystems. The interviews were made in Estonian and translated to English by the interviewing researcher. For detailed methodology, please see Olesk (2019b).

students after graduation. The contacts with other research teams were initiated by students and facilitated by advising researchers. The academic network ranged from the United States to India, and especially close ties in Europe were established with universities from Finland and Denmark. The main collaboration with industry concerned the development of the mission control software. Collaboration with the software company CGI (formerly Logica) included mentoring of the students by the company specialists and involving them in the development process.

The tangible impacts of these engagement strategies and collaborations include the fact that ESTCube-1 was completed and the project continues with ESTCube-2 and ESTCube-3 currently in development. Members of the ESTCube team also developed the camera system for ESA's European Student Earth Orbiter (ESEO) mission. In total, over 30 bachelor's theses and over 20 master's theses were defended in connection with the project. Fourteen scientific articles were published, most having students as coauthors. In addition, several project alumni later founded start-up companies or were employed by technology companies.

ESTCube-1 also involved younger students in the project and related activities. In 2012, the project supervisor initiated "Teadusmalev" (Science Task Force), a yearly summer work camp for talented high-school students. The camp applies the same problem-based learning approach that characterizes the whole project, presenting the students with a scientific or engineering problem related to ESTCube and guiding them to solve the task. Tasks tackled in the camps over the years have included testing the satellite's attitude control system, calculating the spacecraft's trajectory, and building a Helmholtz coil ("Science task force," n.d.). In 2008, the ESTCube project supervisor also initiated a weekly space club for primary school pupils. The club, called Estronauts, allows children to visit science labs and try out scientific equipment, conduct experiments, and learn about various sciences related to space, from astronomy to social sciences.

## Engagement With Media

ESTCube-1 was visible in the Estonian media during the whole project, generating at least 160 original media items (i.e., items that resulted from an interaction between a journalist and one or more team members), a special issue in *Horisont* (the oldest popular science magazine in Estonia), and a documentary film.[2] The media coverage was overwhelmingly positive, and the project was reported as a success even when it became clear that its scientific mission had failed (Olesk, 2019a).

The amount and tone of media coverage can be considered a result of conscious media work by the ESTCube team. The awareness of media's role and potential for

---

[2] *Kuidas ehitada kosmoselaeva?* (*How to build a space ship?*, 2015), directed by Madis Ligema, 75 min.

public engagement, however, did not exist in the team at the beginning of the project. The team members had no or very little media experience and did not initially consider media as part of their activities. This only changed when they issued their first press release announcing the start of the project. While the aim of the press release was to notify other students about the project and invite them to join the team, it also generated media attention for which the team was not prepared. For example, after the first press release, many journalists asked the team what the satellite is going to do in space. The fact that the team had not chosen a mission yet diminished their credibility, the team members reflect in retrospect. They describe their first ideas about communication as amateurish and behavior as trial-and-error.

*This was all learning by doing. We did not know the result but tried. During my first experiences there was a lot of nervousness and rapid response behavior. (Project manager)*

While the media reports about the project were positive, the team found that the majority (76%) of the comments readers had posted online by the news articles about ESTCube-1 were negative (Kvell & Noorma, 2010). Commenters opined that the satellite project was considered a waste of money or a joke, thought Estonia was too small or poor for a space project or displayed ignorance about the project. The experience of not being satisfied with the initial media interactions and public response triggered the initiative to consciously work on improving media skills of all team members and using media to increase public awareness and support for the project. As a result, the team quickly became skillful communicators. This process was supported by the interplay of three elements: the role of the group leader, participation in media trainings, and regular interactions with the media (Olesk, 2019b). These elements helped the team members to understand how media operate, what journalists expect from sources, and what is the proper style to present science for the public.

The project supervisor was the main person to interact with the media. He featured most prominently in media coverage (Olesk, 2019a) and took a proactive role by contacting journalists directly and offering them news about the project. Several team members said that the supervisor's ideas about the importance of public engagement and his presentation style inspired them to engage more actively in public communication.

At the same time, involving junior team members in communication activities was a conscious policy of the team management. They were provided opportunities to give interviews to the media and to give public presentations. This was done to decrease the load of the supervisor and also to boost the credibility of the message:

*When we already had more students, it became clear that any communication, whether a press conference or recruitment of new students, should be done by immediate actors. There is no point of having a spokesperson. People who are doing things must do the talking. (Project manager)*

The project did not assign anyone officially to manage media relations. The communication responsibilities (such as writing press releases or giving interviews) were often distributed among the members on an ad hoc basis. Decisions such as timing and content of press releases and press conferences were often discussed collectively in project meetings. While doing all communication activities on their own, the team occasionally and informally consulted with professionals, mainly journalists and public relations specialists, to learn about design of communication strategies.

The frequent use of media was not part of the project's communication strategy in the beginning. Only when the first press release attracted considerable interest from the media and the team felt that the topic was attractive for the Estonian public did they start to engage with media consciously:

*Small news items, once or twice a year, as we produced them, kept us on the picture…When we approached launch then things started to gear up. This was intentional, of course. (Project manager)*

The goal was to use this visibility to achieve other strategic goals of the project. These ranged from project-specific aims (such as to create a favorable public atmosphere toward the satellite, attract additional funding, and ensure support of the decision makers) to more general aims such as to attract young people to study in STEM fields and promote problem-based learning.

These aims guided the style and messages used in communication. For example, emphasis was placed on using storytelling in articles that the group members wrote themselves and preparing a variety of visuals to be used in print media and TV. Their general aim in public communication was to create positive emotions for the audience. Depending on the channel and its audience, this could have meant using personification of the satellite (e.g., "Our tiny brave satellite is battling with space junk"), statements to incite pride (e.g., "Satellite built by Estonian students tests a revolutionary space travel technology"), or outlining the wider benefits of the project (e.g., "ESTCube supports the growth of the economy via good education"). Perhaps surprisingly, the team paid particular attention to targeting messages to retired people. As they form a vocal group in Estonian society (e.g., in radio call-in shows), the team perceived it to be important to convince them that the satellite project was worthwhile, using the argument that projects such as ESTCube contribute to the national economy, which in order will create a better future for all, including increasing the size of pensions.

ESTCube also attracted the attention of foreign press, receiving coverage by international news agencies such as Reuters and various international publications, among them the popular science magazine *New Scientist*. In this coverage, the main focus was clearly on the nature of the planned scientific experiment, which in domestic media was rather overshadowed by the messages emphasizing the Estonia-related aspects of the project.

During the project, the ESTCube-1 team issued 29 press releases and hosted four major press conferences (announcement of the mission, presentation of the satellite before launch, celebrating 1 year on orbit, and announcement of the end of the mission). At the start, the team used mostly press releases to announce major

**FIG. 9.4**

The ESTCube-1 team and their supporters at the press conference in January 2013 where ESTCube-1 was publicly presented before delivering it to the launch provider. Mart Noorma is in the first row, second from right. *SA Eesti Tudengisatelliit.*

milestones or activities, but toward the end of the project, they shifted to other methods, mostly press conferences and personal contact with journalists. Press conferences were chosen because they give a sense of importance for the event, allow TV to record footage, and give students a chance to practice presenting on stage (Fig. 9.4).

Previous studies (Besley et al., 2013; Entradas & Bauer, 2017; Poliakoff & Webb, 2007) have identified a number of factors that drive scientists to engage with the public, but none of them highlight the intrinsic motivation as strongly as it can be seen for ESTCube. This belief spread during the project from the core management group to the rest of the team. Significant media interest provided ample opportunities for the team to learn and practice the specific skills of how to make the project visible in the media and to use this visibility to achieve the strategic aims of the project. The result was extensive, positive, and supportive media coverage which led, as judged from our subjective viewpoint, to wide public awareness about the project and a similarly positive public discourse.

The skills acquired and lessons learned during engagement with media also helped the team to develop better tools for direct communication with other groups, covered in the next sections.

## Engagement With Decision-Makers

Government decision-makers formed an important stakeholder group who became closely engaged with the project and whose support was crucial for the completion of the project. We distinguish two subgroups of decision-makers: public administration (civil servants working for government ministries and agencies) and elected policy-makers (members of parliament and cabinet).

Soon after ESTCube was initiated, Estonia started the process of joining the European Space Agency (ESA). This process was led in Estonia by the Ministry of Economic Affairs and Communications and the Estonian Space Office, located in the government agency Enterprise Estonia. ESTCube and Estonia's process of joining ESA had started independently of each other, but the ESTCube team and the government organizations soon realized that they shared a number of communication goals, such as convincing the elected policy-makers and the public that Estonia could and should be involved in space affairs.

The primary goal of the Estonian Space Office was to secure the decision to join ESA for which a decision by the parliament was needed. ESTCube also needed political action—as the very first spacecraft of Estonia, legislative changes were necessary to help ESTCube comply with all international spaceflight regulations.

Although these decisions were to be taken by policy-makers, both the Estonian Space Office and ESTCube relied on the assumption that political decisions are easier to make once the decision-makers perceive wide public support for the decision (or at least not resistance against it). Hence, they made efforts to create a favorable public discourse, mostly via media. ESTCube's media work was covered previously. The Estonian Space Office's public activities included the establishment of a website with space-related news and commissioning a TV series about Estonia and space. By outlining the potential of space projects for the Estonian economy and presenting examples of how Estonia is already involved in international space projects, they aimed to prove that the country is not too small to be dealing with space matters (as has been a frequent argument by the critics) and that active involvement in such projects (including via the membership in ESA) could bring economic benefits. ESTCube gained a prominent place in such communication as a good example of Estonia's capabilities in space matters. ESTCube often emphasized in its communication that part of its mission was to educate future professionals for the Estonian space industry. Therefore, the two groups mutually amplified similar messages in the public.

ESTCube also engaged with decision-makers directly, seeking to engage them in project activities and finding opportunities to lobby for the project. Contacts on the personal level with various decision-makers were helped by the smallness of Estonia: it is relatively easy to meet even Cabinet members on various occasions or establish contact with them. The first public presentation of the satellite project was made in collaboration with the Estonian Foreign Ministry, the audience being Estonian diplomats and people acting as honorary consuls of Estonia in various countries. The satellite team also established a good contact with Ene Ergma, an astrophysicist-turned-politician who acted as the speaker of the Parliament in 2003—06 and in 2007—14. She voiced strong support for the project and participated in several ESTCube public events, becoming the third-most-often quoted person in ESTCube media coverage (after project supervisor Mart Noorma and project manager Silver Lätt; Olesk, 2019a).

In addition to the legislative changes, political support became crucial for the project in negotiations to find a launch vehicle. Involvement of decision-makers in discussions with ESA helped to ensure ESTCube-1 a place on the Vega launcher.

We saw that the decision-makers adopted the narrative presented by ESTCube in their public discourse. It became part of the narrative of Estonia as a tech-savvy and innovative country (to the global audience) and a matter of national pride (for the Estonian audience). The satellite project was mentioned by Prime Minister Andrus Ansip in his annual overview of R&D policy in 2012, and the satellite was featured on the official Christmas card of the president in 2013.

## Engagement With the General Public

The Estonian public was one of the key target groups for the ESTCube team. All segments of the population were considered relevant, from the young to the old. On one hand, the messages to the public served to promote science and attract young people to study science, technology, engineering, and mathematics (STEM) by showing that even the most complicated systems, like a spacecraft, could be developed in their homeland. This aim harmonizes with the strong focus on STEM education and career options that have dominated the Estonian science communication landscape (Olesk, 2020).

On the other hand, to build wide support for the project, the team greatly emphasized the wider impacts of the ESTCube project to society, including the economic benefits that could come with the development of the space industry. Supporting both goals was the discourse of national pride, convincing the public that Estonia had both the ambition and talent to undertake significant space projects.

The general public was mostly targeted via media as described previously. Predominant forms of direct interaction were public talks and presentations. Especially after the project supervisor appeared in TEDxTallinn in 2010, members of the team were invited to talk to a wide variety of audiences to introduce the project. As such communication formats were one of the most time-consuming for the team, they were not able to respond to all requests.

In social media, ESTCube is active in Facebook (approximately 6600 followers in 2019) and Twitter (approximately 370 followers in 2019). These channels are consciously aimed at the general public, keeping the posts light on technical details. For more technical descriptions—for which the ESTCube-1 team believed there was also an audience—the team established a web forum. In all channels, the team communication is both in Estonian and in English, keeping in mind the potential international audience, such as other teams building similar CubeSats.

The team occasionally made use of the possibilities of interaction that social media platforms like Facebook provide, for example, engaging users with games (e.g., guess the place on the satellite's photo). Tracking the level of user interaction (e.g., likes, comments, and shares) was considered important for regular posts as well, and the team members aimed to design the posts to increase such interactions. One team member who posted frequently commented: "[Preparing Facebook posts] *was tough. You had to rely on your gut feeling about what worked and what did not. Eventually I got a sense and followed it.*"

Once the satellite had launched, the team occasionally devised special activities for the public. The most visible campaign was on Valentine's Day in 2015. The team opened a website where anyone could send a message to their loved one. A copy of the message was uploaded to the memory chip of ESTCube, and the key concept of the campaign was that people's wishes would continue circling the Earth for years to come. The campaign resulted in a couple hundred messages being sent. Another campaign took place in 2017 by the ESTCube-2 team in collaboration with the Estonian Song Festival organizers, launching a public vote to choose an Estonian song to be uploaded in future satellites. That same year, ESTCube-2 launched a campaign in Hooandja (the Estonian version of Kickstarter) to collect funds for the development of the satellite. The campaign was successful, surpassing the initial 30,000€ target and gathering 38,743€. The team attributes the success in part to strong media presence and active social media engagement (Kalnina et al., 2018).

## Discussion

ESTCube-1 presents us with a case in space science where engagement activities were just as important as the scientific and engineering content of the mission. Likewise, the engagement goals were no less ambitious than the scientific ones. That is, the team defined as goals of the project to develop and build the satellite with student involvement only, to make space science and other STEM subjects more attractive for young people, and to induce positive public attitudes toward Estonian science. This allowed the team to announce that by launching the satellite, they had already accomplished 90% of the mission. At the end of the mission, despite the malfunction of the scientific experiment, both the team and the media declared the project an overwhelming success. ESTCube is one of the publicly best known science projects in recent Estonian history.

This chapter has described the engagement strategies that allowed one senior researcher to compose a dedicated group of students able to complete the satellite mission and the communication activities of the ESTCube team that contributed to the project's popularity. While a good relationship with the media was at the core of the wide public awareness and support, the team was active with various target groups using different engagement methods. ESTCube can be considered noteworthy as a science communication example for mainly two reasons: the scope of its activities, including the variety of target groups, and the fact that almost all communication and engagement was done by the team members themselves.

The latter aspect also means that not everything came out perfectly or fully professionally. In the interviews, the team members described the difficulties they had—especially in the beginning—with successful engagement. Some also expressed criticism about the focus or quality of some of the communication activities. These comments, however, reflect the nature of how the team approached communication: as they did with the engineering or scientific issues in their project, they applied the learning-by-doing approach, in which a constant reflection on the activities is a vital part.

One must note that many of the public communication activities described in this chapter represent the classical dissemination or popularization model of science communication, not quite engagement in the sense of what has been termed as Public Engagement in Science and Technology ("From PUS to PEST," 2002). Beyond the involvement of young people because it was an essential part of the project ideology of problem-based learning, wider public engagement in the technical aspects of the project was not considered in the beginning of the project. The skills and ideas for engagement came during the project but, by that time, the satellite design was already in place and only left space for activities such as the Valentine's Day campaign with little relevance to the actual mission. In the interviews, some team members said that if they would have to start all over, they would consider including some public engagement opportunities relating to the design of the satellite.

Currently, ESTCube-2 and 3 are in development. Although much of the original team that made ESTCube-1 possible has moved on, the appreciation for the value of public engagement is still high in the team.

Would ESTCube be a good model for similar (space) science teams looking to design their engagement activities? On one hand, there were some specifics of the project that are hard to copy, such as the personality and the skills of the project supervisor that became the driving force behind many of the activities. Also, specific circumstances of Estonia such as good access to decision-makers and journalists and the novelty of ESTCube being the country's first-ever satellite helped to make the project as visible and prominent as it became.

On the other hand, we see evidence that one team of researchers (or students) can initiate and manage successful engagement activities, if driven by the understanding that public engagement is an important part of their work. We also see that relevant skills can be learned.

# References

Besley, J. C., Oh, S. H., & Nisbet, M. (2013). Predicting scientists' participation in public life. *Public Understanding of Science, 22*(8), 971–987. https://doi.org/10.1177/0963662512459315

Entradas, M., & Bauer, M. M. (2017). Mobilisation for public engagement: Benchmarking the practices of research institutes. *Public Understanding of Science, 26*(7), 771–788. https://doi.org/10.1177/0963662516633834

Envall, J., Janhunen, P., Toivanen, P., Pajusalu, M., Ilbis, E., Kalde, J., Averin, M., Kuuste, H., Laizans, K., Allik, V., Rauhala, T., Seppänen, H., Kiprich, S., Ukkonen, J., Haeggström, E., Kalvas, T., Tarvainen, O., Kauppinen, J., Nuottajärvi, A., & Koivisto, H. (2014). E-sail test payload of the ESTCube-1 nanosatellite. *Proceedings of the Estonian Academy of Sciences, 63*(2S), 210–221. https://doi.org/10.3176/proc.2014.2S.02

ESTCube-1. (n.d.). Retrieved from https://www.estcube.eu/en/estcube-1.

From PUS to PEST. (2002). *Science, 298*(5591), 2–49. https://doi.org/10.1126/science.298.5591.49b

Hmelo-Silver, C. E. (2004). Problem-based learning: What and how do students learn? *Educational Psychology Review, 16*(3), 235–266. Retrieved from http://kanagawa.lti.cs.cmu.edu/olcts09/sites/default/files/Hmelo-Silver_2004.pdf.

Janhunen, P. (2004). Electric sail for spacecraft propulsion. *Journal of Propulsion and Power, 20*(4), 763–764. https://doi.org/10.2514/1.8580

Kalnina, K., Bussov, K., Ehrpais, H., Teppo, T., Kask, S-K., Jauk, M., Slavinskis, A., & Envall, J. (2018). Crowdfunding for satellite development: ESTCube-2 case. In *IEEE aerospace conference proceedings* (Vol. 2018, pp. 1–14). https://doi.org/10.1109/AERO.2018.8396722

Kvell, U., & Noorma, M. (2010). *Social impact of Estonian student satellite program.* Retrieved from http://www2.isunet.edu/index2.php?option=com_docman&task=doc_view&gid=2373&Itemid=26.

Lätt, S., Slavinskis, A., Ilbis, E., Kvell, U., Voormansik, K., Kulu, E., Pajusalu, M., Kuuste, H., Sünter, I., Eenmäe, T., Laizans, K., Zalite, K., Vendt, R., Piepenbrock, J., Ansko, I., Leitu, A., Vahter, A., Agu, A., Eilonen, E., … Noorma, M. (2014). ESTCube-1 nanosatellite for electric solar wind sail in-orbit technology demonstration. *Proceedings of the Estonian Academy of Sciences, 63*(2S), 200–209. https://doi.org/10.3176/proc.2014.2S.01

Olesk, A. (2020). Estonia: Science communication in a post-Soviet country. In T. Gascoigne, B. Schiele, J. Leach, M. Riedlinger, B. V. Lewenstein, L. Massarani, & P. Broks (Eds.), *Communicating science: A global perspective* (pp. 279–296). ANU Press.

Olesk, A. (2019). Media coverage of a strongly mediatized research project: The case of the Estonian satellite ESTCube-1. *Mediální Studia, 13*(1), 7–27.

Olesk, A. (2019). Mediatization of a research group: The Estonian student satellite ESTCube-1. *Science Communication, 41*(2), 196–221. https://doi.org/10.1177/1075547018824102

Poliakoff, E., & Webb, T. L. (2007). What factors predict scientists' intentions to participate in public engagement of science activities? *Science Communication, 29*(2), 242–263. https://doi.org/10.1177/1075547007308009

Science Task Force. (n.d.). Retrieved from https://www.estcube.eu/en/the-science-task-force.

Slavinskis, A., Pajusalu, M., Kuuste, H., Ilbis, E., Eenmäe, T., Sünter, I., Laizans, K., Ehrpais, H., Liias, P., Kulu, E., Viru, J., Kalde, J., Kvell, U., Kütt, J., Zalite, K., Kahn, K., Lätt, S., Envall, J., Toivanen, P., … Noorma, M. (2015). ESTCube-1 in-orbit experience and lessons learned. *IEEE Aerospace and Electronic Systems Magazine, 30*(8), 12–22.

Slavinskis, A., Reinkubjas, K., Kahn, K., Ehrpais, H., Kalnina, K., Kulu, E., Pajusalu, M., Kvell, U., Sünter, I., Kuuste, H., Allik, V., Lätt, S., Vendt, R., & Noorma, M. (2015). The Estonian student satellite programme: Providing skills for the modern engineering labour market. In *Proceedings of first symposium on space educational activities.*

# From Green Peas to STEVE: Citizen Science Engagement in Space Science

# 10

**Lucy Fortson**

## Introduction

In 1999, the spacecraft Stardust was launched from Cape Canaveral to rendezvous four and a half years later with the comet Wild 2. As the probe approached its target, a robotic arm unsheathed itself, raising and locking into place an awkward-looking catcher's mitt tiled with blocks of Aerogel, a low-density material. Particles emanating from the comet and elsewhere that hit the Aerogel buried themselves within, preserved for further investigation upon their return to Earth. This was the first robotic sample-return mission conducted by NASA and was successfully completed with the 2006 reentry and touchdown of the protectively encapsulated Aerogel in the desert of western Utah. By analyzing the tracks made by the particles that embedded themselves into the Aerogel, along with other features, the Stardust team hoped to learn about the environment and chemical makeup of the comet, potentially shedding light on solar system formation theories. However, analyzing the tracks required millions of digital "slices" of the Aerogel tiles to be generated and the track locations to be identified within the slices. The team deployed an unusual strategy to solve this issue: they developed the Stardust@Home website with a "virtual microscope" and invited the public to help analyze the dust tracks. This site was launched in 2006 and over 20,000 volunteers passed the established training regimen, eventually searching over 46 million images in the first year (Westphal, Anderson et al., 2014). By 2014, the Stardust team had several publications, with over 30,000 volunteers contributing to results including the discovery of the first known interstellar dust particle captured (Westphal, Stroud et al., 2014). As of 2021, Stardust@Home is still going strong along with its place in history. While other projects in the early 2000s, such as Mars Clickworkers (Coleman et al., 2014), laid the groundwork for engaging volunteers through a distributed Internet call to action, Stardust@Home established that web-based citizen science could draw a crowd and, crucially, that crowd could also deliver robust data products, leading to highly cited peer-reviewed research publications. Citizen science as a form of online public engagement in space sciences took one giant step forward with Stardust@Home.

**Space Science and Public Engagement. https://doi.org/10.1016/B978-0-12-817390-9.00009-9**

Chris Lintott and Kevin Schawinski (postdoctoral scholar and graduate student, respectively, at the University of Oxford in 2007) had their own "big data" problem in confronting the challenge of producing a robust catalog of galaxy morphologies from the million or so galaxy images recorded by the Sloan Digital Sky Survey (SDSS). Historically, astronomers have labeled or classified "by eye" whether a galaxy appears in an image as a disk with spiral arms or shaped more like an elliptical blob, a distinction which happens to be important for understanding the evolution of galaxies. Kevin Schawinski had taken an entire week to classify 50,000 of the SDSS galaxies. As the story goes, Chris and Kevin retired to an Oxford pub to work out how to proceed with classifying all the rest without ruining Kevin's life. Having heard about Stardust@Home, they thought it seemed like a good idea to put the galaxy images into a web interface and enlist the help of the public to classify those images. With Stardust@Home participation numbers in mind, they reasoned that it would take about 3 years to obtain sufficient classifications from the public on the remaining SDSS galaxy images. Meanwhile, there was plenty of other science to get done. With this inspiration, they recruited two other Oxford postdocs, Kate Land and Anže Slosar, and got down to the business of gathering the images and building the website.

Galaxy Zoo, as the ensuing project was called, was launched in 2007, and much to the delight of the team, the public response was overwhelming—so much so that within the first few hours, the host webserver was brought to its knees. Team members at Johns Hopkins University quickly changed the server (even though it was the middle of the night) to enable the flood of classifications to continue to pour in and reach 40,000 classifications per hour within 24 h of launch. By September 19, 2007, just 2 months after going live, Galaxy Zoo had 106,806 volunteers and 29,292,826 classifications, averaging about 30 classifications per each of the million SDSS galaxies. This first version of Galaxy Zoo finished in February 2009 with over 80 million classifications by more than 170,000 volunteers—in about half the time for which the project had originally been planned.[1] Meanwhile, the Galaxy Zoo team had an inkling that this was only the beginning of something far greater—that it was not just galaxies people enjoyed classifying—and in which case, developing a platform for citizen science projects made a lot of sense: projects would benefit from a common infrastructure as well as shared access to a volunteer base, and volunteers would benefit from similar user experience designs as well as social aspects of a more cohesive online community. Before building such a citizen science platform, several questions had to be answered first: Was Galaxy Zoo a one-off success or would volunteers come back in the same numbers for a follow-on Galaxy Zoo 2 project? What was the fundamental motivation for volunteers to participate in Galaxy Zoo 1? Would volunteers be willing to contribute to fields outside of astronomy? And how difficult a task would volunteers be willing to do? Furthermore, Galaxy Zoo

---

[1] For a more complete review of the history and science of Galaxy Zoo, see Fortson et al. (2012) and Lintott et al. (2011).

had given rise to several outright serendipitous discoveries by the volunteers, including new types of galaxies, such as the Green Peas (Cardamone et al., 2009), and previously unknown astronomical objects, like Hanny's Voorwerp (Schawinski et al., 2010). How could these discoveries continue to be enabled?

Members of the Galaxy Zoo team did go on to build a platform for citizen science called the Zooniverse (www.zooniverse.org), launched December 12, 2009, and have found answers to many of these questions. Yet, over 10 years later, forms of these same questions remain at the root of all citizen science projects engaging with space science. What has changed is the scale of what we mean by big data and the concomitant introduction of machine learning. For example, Galaxy Zoo was able to tackle the big data scales of the 2000s, but even if every person on Earth takes part in classifying, humans alone would not be able to get through the billions of galaxies in the data sets coming in from the next generation of telescopes such as the Large Synoptic Survey Telescope (LSST) (Ivezi, 2013) - recently renamed the Vera Rubin Observatory (VRO). To tackle this new scale of big data, we must learn how to work alongside artificial intelligence. The last decade has also seen a maturation of the field with the acceptance of citizen science as a methodology which can produce robust data products. Further, the development of sensors and the evolution of web-based tools has enabled the proliferation of citizen science in serious research projects as well as a means for improving science education. There has also been excellent research on the science of citizen science (for example: how do contributors stay engaged? do participants learn science?) and the organization of citizen science into governance bodies at national and international levels has led to good things like community-of-practice conferences and metadata standards. Drawing on my experience as a cofounder of the Zooniverse platform, this chapter will touch on several of these developments over the past decade with the intent of providing a real sense of how citizen science has become a complete game changer in the twin goals of accomplishing research and engaging the public in the space sciences.

## The Rise of Citizen Science

While its history dates back centuries, citizen science has taken its place in the past decades as a highly successful tool to solve some of the most challenging research problems across a wide range of domains (Bowler et al., 2013; Cooper et al., 2010; Garneau et al., 2017; Lintott et al., 2011; Newman, 2014; Sullivan et al., 2009; Theobald et al., 2015). Problems that require a distributed call to action are particularly suitable, leading to the three major types of citizen science: (1) the collection of spatiotemporally distributed data, as in the Community Collaborative Rain, Hail and Snow Network (CoCoRaHS) network of rain, hail, and snow gauges monitored by volunteers across the country and used in data products created by US government agencies ranging from NASA to the National Oceanic and Atmospheric Administration (NOAA) to the US Department of Agriculture (USDA) (Reges et al., 2016); (2) as a form of distributed computing where the public donates spare computer cycles, as in Einstein@Home's search for spinning neutron stars with data from the Laser Interferometer Gravitational-Wave Observatory (LIGO) gravitational wave detector (Abbott et al., 2009); and (3) the analysis by multiple independent classifiers of large, complex data sets, such as images of galaxies whose morphological characteristics are labeled in the Galaxy Zoo project and provided as value-added data products to the image surveys (Lintott et al., 2011; Willett et al., 2015, 2017). Additionally, (Bonney, Ballard, et al., 2009) describes three levels in terms of volunteer engagement which are (1) contributory, in which volunteers perform a well-defined task designed by a research team such as collecting data; (2) collaborative, in which volunteers may contribute beyond simple task performance (for example, by suggesting changes to protocols or participating in analysis of the data); and (3) cocreated, in which volunteers work with a research team up front in designing and implementing a citizen science project.

It is hard to pinpoint exactly which snowflake started the avalanche of citizen science, but it is pretty clear that the rate of snowfall was beginning to increase in the late 20th century and increased dramatically in the first two decades of the new millennium (see Cappadonna et al. (2017) for a brief history of citizen science and the use of the term). Follett and Strezov (2015) describe a comprehensive study of the publication databases Scopus and Web of Science where the authors filtered peer-reviewed publications using the term "citizen science," finding a total of 1127 unique articles (see Fig. 10.1). They note that the number of articles related to citizen

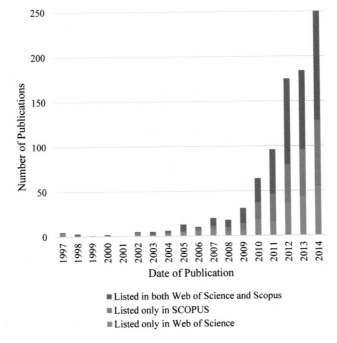

**FIG. 10.1**

Growth over time of peer-reviewed articles in both Web of Science and Scopus containing the term "citizen science."

*Reprinted from Figure 3, Follett, R., & Strezov, V. (2015). An analysis of citizen science based research: Usage and publication patterns. PloS One, 10(11), e0143687, through Creative Commons License.*

science should be taken as a lower bound, as many relevant articles may use terms other than, but related to, citizen science (such as "crowdsourcing") or may not acknowledge the use of data obtained through citizen science methods (see Cooper et al. (2014) for further discussion on the difficulties of determining which articles use data collected via citizen science). Nonetheless, the explosive growth in articles using citizen science can be seen in Fig. 10.1.

While the modern adoption of citizen science probably owes its roots to the environmental activism of the 1960s kicked off by Rachel Carson's *Silent Spring*, there are a number of intertwining factors that likely played a role in the recent phenomenal rise of citizen science. Though I know of no formal study that substantiates this claim, it seems to me that the biggest factor has to be the rise of the Internet in general and the World Wide Web in particular (including, most recently, social media platforms). The concomitant increase in the sheer amount of data being produced through new sensor technologies and then made available through the Internet also has had a significant impact on citizen science. It is this dispersive nature of the Internet that allows for both a larger-scale call to action and the ability to distribute data to and collect data from those who take up the challenge. These factors directly contributed to the development of the second and third citizen science types mentioned earlier, namely online data analysis through either distributed computing or participation in performing analysis tasks distributed via the Internet. Furthermore, the distributed nature of the Internet improved the reach of data collection citizen science, and web interfaces were developed for many venerable data collection projects such as the Christmas Bird Count (see, for example, Murthy et al. (2016)) to help improve data entry and data quality metrics (see, for example, Sullivan et al. (2014), Wiggins et al. (2011)). Finally, the Web and social media have brought together scattered groups of like-minded people to focus on specific interest areas. Distributed communication enabled by the Internet provides the oxygen for the growth of citizen science platforms such as Citsci.org (Newman et al., 2011), CosmoQuest (www.cosmoquest.org) (see, for example Robbins et al. (2014)), iNaturalist (www.inanturalist.org) (see, for example, Mason Heberling and Isaac (2018)), SciStarter (www.scistarter.org) (see, for example, Hoffman et al. (2017)), and the Zooniverse (Fortson et al., 2012; Trouille et al., 2019). Without blogs, tweets, Facebook posts, and even email, these platforms would not survive to engage participants in a wide array of projects.[2]

Another set of factors likely contributing to the rise of citizen science lies in the sea change over the past few decades in attitudes surrounding the public engagement in science. In 1995, NASA initiated an entire space science education and public

---

[2] Of course, very meaningful citizen science is still done the "old-fashioned" way through physical meetings of communities of volunteers and researchers, in particular where the tasks require high levels of training for best outcomes (see, for example, the Coastal Observation and Seabird Survey Team (COASST) project (Parrish et al., 2017) and other location-based monitoring projects that thrive using pencil and paper (Danielsen et al., 2005)).

outreach (EPO) ecosystem through implementing *Partners in Education: A Strategy for Integrating Education and Public Outreach into NASA's Space Science Programs*, which eventually required EPO funding to be part of researcher-proposed science missions and provided for separate, smaller EPO funding enhancements for individual science grants (Rosendhal et al., 2004). Among many other aspects of the EPO initiative, this led to a large increase in the number of scientists participating in nontrivial efforts to engage the public in the understanding of space science (Smith et al., 2014). In a 2012 Astronomical Society of the Pacific conference report entitled *Barriers, Lessons Learned, and Best Practices in Engaging Scientists in Education and Public Outreach*, it is explicitly shown that funding from the NASA EPO ecosystem alleviated one of the key barriers mentioned: lack of ability to pay scientists for their time to take part in EPO efforts which then minimizes their motivation to do so (Buxner et al., 2012).

There is also evidence that the injection of tens of millions of dollars annually by NASA into its EPO initiative literally created the "EPO professional" and a new career pathway for young scientists to pursue (Fraknoi, 2005). Dang and Russo (2015) reported in a study of 155 professional astronomers an overwhelmingly positive attitude toward the need for EPO. As we will see in the following, this was a critical step in setting the stage for the adoption and proliferation of citizen science in space sciences.

In addition to NASA's efforts, in 1997, the National Science Foundation (NSF) introduced as a separate proposal review criterion the concept of "broader impacts." In a recent review, the executive summary states:

> *Researchers are increasingly reaching out to the general public as a means of raising awareness and increasing appreciation of the role that science plays in the quality of everyday life. The spread of new media tools has only increased the opportunities for scientists and educators to interact with the public.*[3]

Both the NASA and NSF initiatives are fundamentally driven by the many congressional and National Academies reports that point to a real need to improve the STEM education pipeline and engage the public in science (see, for example, the America COMPETES Act). Compelling scientists through funding sources to participate in such a critical national effort makes the most sense when the public engagement activities the scientists pursue are closest to their research (Thiry et al., 2008). As a new generation of scientists were primed through the 1990s and 2000s with the massive EPO push described earlier to be receptive to engaging the public in science, the option of developing a research project that engaged the public such as through citizen science was a seed falling on fertile ground. This point was not lost on individuals in federal science agencies, including NASA, who

---

[3] See https://www.nsf.gov/od/oia/publications/Broader_Impacts.pdf.

formed an interagency working group to promote citizen science as a research method to be used by agency scientists and those scientists funded by the agencies. Effort was required to understand the legal aspects of federal agencies working with, for example, the complex intellectual property rights that can emerge through citizen science projects.[4] This work led to a 2015 White House Office of Science and Technology Policy memorandum entitled "Addressing Societal and Scientific Challenges through Citizen Science and Crowdsourcing" that has underpinned the burgeoning adoption of citizen science across federal agencies and an effort to coordinate federal activities through agency-designated citizen science and crowdsourcing coordinators[5] as well as the Federal Community of Practice for Crowdsourcing and Citizen Science (FedCCS),[6] which now encompasses over 350 members across at least 60 agencies. In addition, state agencies, especially many Departments of Natural Resources, have embraced citizen science to help tackle the ever-growing task of environmental monitoring.

In addition to the adoption of citizen science as a tool by a wide range of US state and federal agencies, the growth in the sheer number of researchers, educators, managers, and others involved with citizen science necessitated the establishment of professional societies in the United States as well as in Europe and Australia. In the United States, the Citizen Science Association (CSA) has now held three national conferences since 2015, growing from 600 participants in 2015 to over 800 in 2019; in May 2016, the CSA's academic journal *Citizen Science: Theory and Practice* was launched. Working groups deal with a wide range of common issues such as metadata standards and ethics, professional development, and policy. However, it is important to note that it is a very difficult task to satisfy the needs and address the concerns of all disciplines active in citizen science. While citizen science related to space science is represented throughout the CSA membership, there is certainly room for space science teams involved in citizen science to convene their own meetings and working groups to address specific concerns.

## Metrics of Success: Data Quality and Publications

Since its modern inception, citizen science has wrestled with concerns raised by professional scientists that it is not a legitimate method of producing publishable research products. This lack of legitimacy seems to be derived directly from the very name "citizen science:" by equating the term citizen with public, their

---

[4] See, e.g., https://www.wilsoncenter.org/sites/default/files/research_brief_guide_for_researchers.pdf.
[5] https://www.citizenscience.gov/agency-community.
[6] https://digital.gov/communities/crowdsourcing-and-citizen-science/.

reasoning goes, we must be talking about science that is accessible to the public and therefore not the act of producing science.[7]

The argument typically goes as follows: Professional scientists spend years becoming experts in their specific fields, climbing a mountain of literature and honing often highly nuanced research techniques that enable them to finally publish "original research." How would it be possible for the average citizen to contribute to meaningful research if they had not followed the same path? But this question focuses only on the end goal of producing "experts," assuming that somehow the world is divided into "those that produce science" and "those that consume science," and citizens (aka the general public) are definitely categorized on the consuming side of that divide. There is some acknowledgment of "amateurs," and interestingly two of the areas in which citizen science has taken root—astronomy and ornithology—have long histories of amateur contributions. But generally, in this divided world of science producers and consumers, there is no room for allowing agency to develop in a "consumer," which is critical for real engagement in understanding and appreciation of science (Arnold & Clarke, 2014). So we have a false dichotomy: citizens cannot be scientists, and therefore, citizen science must solely have educational or outreach value at best. At worst, to some members of the scientific establishment, it is a sham that purports to enable nonexperts to conduct original research.

To resolve this false dichotomy, we must move away from the "who" to the "what." The crucial distinction is that citizen science is a method of conducting research and thus subject to all the rigors and constraints required of any research method. As a tool, it must be used only where appropriate. It must be calibrated and the ensuing data cleaned and debiased. Results obtained through employing the citizen science methodology must be replicable. This is no different than any other scientific instrument. What is different is that the "pixels" or "information channels" that form the citizen science instrument are made up of human brains in all their complex subtleties. So, in fact, the "what" and the "who" are inextricably linked, and the false dichotomy can only be resolved if we acknowledge that engaging with the process of science by members of the general public can lead to a softening of the divide between science producers and consumers.

Citizen science is thus an "on ramp" to legitimate production of original research. Of course, few participants would have the expertise required to carry out a full research experiment; citizen science is really a "collective" expertise enabled through those who contribute at "entry"-level tasks (motivated to do so because they feel it allows them to contribute to something greater than themselves), those who contribute at more complex tasks, and the few who take on their own

---

[7] There is currently an important debate as to whether the term "citizen" should be replaced by something else as it may be too riddled with negative connotations. This debate is important in the context of citizen science as a form of environmental justice where the term "community science" may be more appropriate. Here, I take what I believe to be the consensus on the original meaning in which a citizen is a member of the general public.

research agendas using available data. Citizen science is the ultimate scientific collaboration that should not threaten the position of professional scientists but should acknowledge that each collaborator contributes to the best of their ability, training, and capacity, as motivated by differing reasons. That this collective intelligence results in peer-reviewed papers—the coin of the academic realm—is the proof in the proverbial pudding of its legitimacy as a method of scientific inquiry.

Indeed, the rise in the number of publications using data from the citizen science methodology (see Fig. 10.1), combined with the strong endorsement by a range of US government agencies for their scientists to adopt citizen science as a tool when and where applicable, provides strong evidence that the scientific community at some level has gotten over concerns that citizen science does not produce quality data. Data quality issues with data collection citizen science can be particularly pernicious as this form of citizen science relies on self-reporting from the volunteers. However, citizen science practitioners have invested considerable thought and effort to ensure data quality and reliability. Here, the careful design and testing of protocols whether online or "in-hand" along with training on how to correctly use them has been shown to produce high-quality data (Bird et al., 2014; Freitag et al., 2016; Sullivan et al., 2009; Wiggins et al., 2011). For example, information campaigns such as "zero counts" have been mounted in monitoring projects to make it clear that reporting zero sightings for a specific species at well-defined repeated observing intervals at a specific location is important to the overall science objectives of the project (Johnston et al., 2019).

For the online data analysis approach to citizen science taken by the Zooniverse, data quality is inherently tied to the production of consensus results based on repeated classification of the same image by multiple independent classifiers. Over the years, Zooniverse and partnering research teams have developed a range of techniques to aggregate classifications into a consensus measure. These can be straightforward, such as simple averaging or weighting algorithms (see, for example, Willett et al. (2017)), or rather more sophisticated techniques where volunteer responses are calibrated against "gold standard" data allowing for individual volunteer experience and skill to be taken into account (see, for example, Marshall et al. (2016), Simpson et al. (2013)). These latter algorithms can also be used when deciding on whether to declare an image "retired" from the system or whether it needs further classifications by other volunteers. This means images can be retired dynamically, requiring, say, only three votes from highly skilled volunteers for a galaxy, which is obviously a spiral as opposed to the static retirement rules from the original Galaxy Zoo era, which required 40 classifications no matter how easy the galaxy was to classify. There are real subtleties involved in removing, or at least understanding, any potential biases inherent in the consensus results. A quick example will help to elucidate this issue. Imagine a galaxy image which looks to the majority of participants on Galaxy Zoo as having no features or spiral arms; the consensus result would be "elliptical." However, this could be an image of a spiral galaxy that is far enough away from Earth that its features appear unresolved by the SDSS imaging system; the consensus result would then be incorrect. There are several ways in which this bias can be taken into account (see, for example, Willett et al. (2017)), but this only works up to a point.

Thus, as with any other instrument, it is important for researchers working with data produced via citizen science in any of its modes to carefully assess the impact that biases will have on pursuing a specific line of inquiry. There are now many papers comparing data produced via citizen science with "expert" data, and when proper citizen science project design and implementation protocols have been employed, the comparisons are generally favorable (Lintott et al., 2008; Swanson et al., 2015).

As citizen science has grown over the past several decades, so too have centralized platforms that make use of the Internet for ease in depositing data collected via smart phones (such as iNaturalist) or online analysis of images (such as Zooniverse). These platforms tend to be operated through groups with several members active in domain research located in academic departments or nonprofits such as museums where research is supported. The support of these platforms, which require professional web development and data science positions, is typically through grant programs at the federal, state, and local level or through private foundations. As citizen science techniques become more mainstream and are adopted by an increasing number of researchers as part of their toolkit, it is important for funding agencies to recognize the need to support the maintenance of, not just the research with, this critical infrastructure.

## Metrics of Success: Discoveries Made by Citizen Scientists

By engaging the volunteers in the primary task for a project, researchers working with Zooniverse have established a track record of producing quality data for use by the wider scientific community (see for example, Hennon et al. (2015), Johnson et al. (2015), Kuchner et al. (2017), Lintott et al. (2008), Schwamb et al. (2013), Willett et al., 2017, Zevin et al. (2017)). However, as alluded to above, there are several instances in which volunteers have contributed beyond "just" performing the primary task. Aided by the Zooniverse discussion board "Talk," volunteers are able to post comments on anything they observe about the images or the task and to discuss these observations with other volunteers and the research team. These actions not only have been shown to contribute to increased science literacy (Masters et al., 2016) but in some cases have led to serendipitous discoveries.

The discovery of the Galaxy Zoo Green Pea galaxies is the poster child for the ideal outcome of providing the right tools to volunteers engaging with a primary task on a citizen science project. In the run-up to launching the website, a decision was taken by the Galaxy Zoo team to make available a link on the primary task to the SDSS Object Explorer page associated with the galaxy image being shown. The SDSS Object Explorer page provides a wide range of "at-a-glance" metadata including redshift, filter magnitudes, and spectroscopic plots as well as further links to literature about that object. There was no real reason to believe that any of the volunteers would avail themselves of the information accessible through this link, but the decision to include the link turned out to be critically important in enabling volunteers to carry out their own research agendas. Just as important was the creation of the discussion forum, which occurred 2 weeks after launch as somewhat of a "self-defense" mechanism due to the overwhelming number of emails the team was receiving from volunteers. The first "Green Pea" (see Fig. 10.2A) was posted to the Forum under the humorous thread title "Give Peas a Chance."[8] But once it became clear that other volunteers had seen such objects as well and were posting them to the thread, several volunteers began to try to find out more about what they might be. Using the Object Explorer page, they began to set forth criteria for what was and was not a Pea, noticing in particular that Green Peas had a characteristic spectrum that was quite different from the run-of-the-mill galaxies they had been classifying on the Galaxy Zoo site.

The spectrum of a Pea galaxy turns out to have an unusually high value of the doubly ionized oxygen emission line, which, for the SDSS filters, provides the characteristic green hue to the galaxies. For several months after the first post, the volunteers collected further examples of Green Pea galaxies, found and discussed literature relating to oxygen-rich emission-line galaxies and their relationship to star forming regions, and generally did what they could to establish a sample of these rare objects. About a hundred had been collected by the time the science team really started to pay attention and Yale graduate student Carolin Cardamone began to work on the scientific interpretation of the Green Peas. These hundred Peas formed the basis of the population search across the entire SDSS, yielding a sample of 251

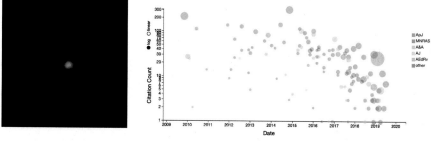

FIG. 10.2

(A) (left): The first Green Pea galaxy to be posted on the Galaxy Zoo Talk Forum "Give Peas a Chance." Credit: Sloan Digital Sky Survey; permission through Creative Commons CC-BY license). (B) (right): Number of citations versus time for articles on NASA's Astrophysics Data System (ADS) filtered by a full text search for the term "green pea." The circle size corresponds to the "read count," or number of reads within 90 days. The color corresponds to the journal in which a given article was published (created by the author courtesy of NASA ADS Explore feature).

---

[8] http://www.galaxyzooforum.org/index.php?topic=3638.0.

Pea candidates, which were shown to be an unusual class of emission-line galaxies with low mass, low metallicity, and very high star formation rates that are rare in the local universe (Cardamone et al., 2009). Straub (2016) describes the details of the volunteer interactions that led to the discovery of Green Peas as a new class of galaxy and validates their work as discovery research. The scientific legacy of the discovery of the Green Peas can be seen in Fig. 10.2B, which shows the number of citations versus time for the 132 articles in NASA's Astrophysics Data System that contain the search term "green pea" (manually filtered for any article not related to Galaxy Zoo Green Peas). The original publication by Cardamone et al. (2009) generated a veritable cottage industry of scientific interest and follow-up on these objects, which are thought to be local universe analogs of intense star-forming galaxies more common in the early universe.

The Green Peas are but one of the many exciting finds made by Zooniverse volunteers in space science. A nonexhaustive list includes discoveries in solar science (Barnard et al., 2014; Davis et al., 2012), exoplanet candidates and discoveries (Fischer et al., 2012; Schwamb et al., 2013) (including the famous Tabby's star (Boyajian et al., 2016; Boyajian et al., 2018)), debris and protoplanetary disks (Kuchner et al., 2016), galaxy morphology and evolution (Keel, Chojnowski et al., 2012; Keel, Lintott et al., 2012; Lintott et al., 2009; Schawinski et al., 2010; Simmons et al., 2014), radio galaxies (Banfield et al., 2016), and gravitational lenses (Geach et al., 2015; Küng et al., 2015; Küng et al., 2018; Marshall et al., 2016; More et al., 2016). Each of these peer-reviewed papers lists volunteers as a member of the author list. In addition, many of these papers also list members of the Zooniverse software and infrastructure development team to credit their contributions as well.

On the data collection side, one of the most remarkable success stories of citizen science contributions to space science is in the discovery of STEVE (Strong Thermal Emission Velocity Enhancement)[9] in images of aurorae taken by volunteers and in some cases posted through the Aurorasaurus citizen science project (Kosar et al., 2018). Apparently known to amateur auroral photographers for decades, STEVE has only recently become a focus of efforts by research scientists to understand its origin. Given its bright colors, the awe-inspiring celestial light display would naturally be associated with the particle precipitation phenomena that characterize the typical green auroral glow when high-energy electrons streaming in from space excite atmospheric oxygen (see Fig. 10.3). However, recent publications indicate that STEVE is actually associated with a hot, high-velocity plasma stream only 50 km wide in the north—south direction but thousands of kilometers wide in the

---

[9] The name STEVE was originally chosen by the volunteers from the Alberta Aurora Chasers Facebook group after a scene in the 2006 movie *Over the Hedge* in which a main character dramatically suggests calling an unknown object "Steve." The current name is an example of a "backronym" attributed to aurora scientist Robert Lysack (see http://www.startribune.com/meet-steve-a-sky-phenomenon-coming-into-its-own/469002563/).

**FIG. 10.3**

Image of the STEVE phenomenon recorded by Rocky Raybell on May 7, 2016, showing both the purplish color typical of STEVE and the green "picket fence" structure that can accompany STEVE. *Rocky Raybell, Creative Commons License Link to license terms: https//creativecommons.org/licenses/by/2.0/.*

*Credit: Rocky Raybell, Creative Commons License Link to license terms: https//creativecommons.org/licenses/by/2.0/.*

east—west direction (Gallardo-Lacourt et al., 2018; MacDonald et al., 2018). The distinctive purplish color is also not associated with the more common electron-induced aurora. So while scientists have not yet determined the exact nature of STEVE, they now know that it is not technically an auroral phenomenon but rather a previously unknown atmospheric effect coincident with, but separate from, aurorae.

As mentioned earlier, citizen science is really a collective intelligence where each individual contributes according to their capacity and interest. Therefore, in the most ideal case, citizen science projects would always offer the opportunity for cocreation of research, in the framing of Bonney, Cooper et al. (2009). This is a time-intensive undertaking by all involved and the reality is that not many projects reach the level of cocreation. But it is imperative that those in the citizen science profession not lose sight of this ideal and work to deploy tools for investigating and visualizing project-related data as these tools become more ubiquitous and accessible. As an example of how citizen science team leads can be mindful of enabling multiple paths of participation in the scientific process, Marshall et al. (2015) invited volunteers from across a range of space science projects in Zooniverse to contribute to a manuscript in *Annual Reviews of Astronomy and Astrophysics*, which described the status of research using citizen science in space science as of 2014. About 15 volunteers participated.

Making the hidden treasures in data more accessible to volunteers is becoming even more crucial as we deploy machine intelligence on citizen science platforms such as the Zooniverse. As described in the following, we risk losing the ability to make serendipitous discoveries such as the Green Peas when machine learning takes on the majority of the classification burden. We must develop a system that enables volunteers to find the weird and wonderful lurking in the data even as the rates of data increase to the point where it is impossible for each image to be seen by human eyes.

## Metrics of Success: Citizen Science Motivation and Engagement

Why would anyone freely give up their time to participate in a citizen science project? How does recruitment and training of participants work? What does a team running a citizen science project need to do to keep participants engaged?

### Motivation

The first question is one of motivation, and this was a question of great importance to the team that had dreams of transitioning Galaxy Zoo to the Zooniverse. Did Galaxy Zoo garner a huge public response because people inherently enjoyed looking at beautiful pictures of galaxies? Were they only interested in astronomy? With a survey of 11,000 volunteers from Galaxy Zoo conducted in 2009, Raddick et al. (2010, 2013) showed that the greatest single motivation for people to participate in Galaxy Zoo was "to contribute to research." Empowered by this result, the Zooniverse team launched the nascent platform in December 2009 with the goal not only to expand into domains beyond space science but also to present more complex tasks. For example, the team asked volunteers to engage directly with data for the Planet Hunters project, where they looked for dips in stellar light curves in data from the Kepler satellite—telltale indicators of the crossing of planets in front of stars (Schwamb et al., 2012).

The Raddick et al. (2010, 2013) study only surveyed Galaxy Zoo participants. A follow-on study surveying volunteers across Zooniverse projects was conducted in

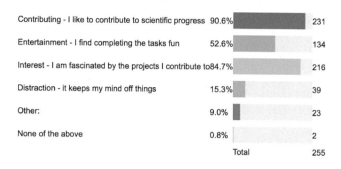

**FIG. 10.4**

Answers to the survey question: "Which of the following words describe your view of participating in the Zooniverse? Please tick all that apply."

*Used with permission from data collected by Homsy, V. (2015). Engagement in the Zooniverse. (Master's thesis, MSc Computer Science). UK: St Cross College, University of Oxford.*

2014 when the platform was host to 37 separate projects. Here, 300 participants were surveyed seeking to gain insights on the makeup of the overall Zooniverse volunteer corps and their perceptions of contributing to the Zooniverse projects. This survey formed the data for a master's thesis by Victoria Homsy (Homsy, 2015), and some of the results were presented in a Zooniverse blog post.[10] Fig. 10.4 shows the results for why respondents participate in Zooniverse projects. A striking 90.6% chose "I like to contribute to scientific progress" as one of their responses. This again underscores what was learned with Raddick et al. and is one of the most important general messages that should be broadcast whenever possible to the scientific establishment as well as policy-makers. The flip side of this message is the key ethical point that volunteer effort must not be wasted and therefore project goals and design must not mislead the volunteers as to the utility for research of the tasks they perform. This means that project managers should never in any way imply that by participating in their citizen science project, volunteers will contribute to research when there is no chance of that occurring. This also means that when a project collects sufficient data or classifications for the immediate research goals, the project should be ended, paused, or relaunched with a clearly messaged educational goal.

## Recruitment and Training

Recruitment strongly depends on the training and engagement expectations placed on the volunteers by the research team and can be highly dependent on the type of citizen science project required. For some projects, there is a strong need to recruit volunteers with specific skill sets (either preexisting or gained through substantial

---

[10]  https://blog.zooniverse.org/2015/03/05/who-are-the-zooniverse-community-we-asked-them/.

training by the team) to perform the tasks required by the project goals. For example, to ensure robust data products, many data collection projects will want to hold in-person training sessions with an expectation of return visits. While some data collection projects are facilitated by online training and data entry, the tasks themselves may be inherently burdensome or time intensive by requiring interaction with the "real" world. For example, the Aurorasaurus project asks volunteers to contribute a record of whether they have seen an aurora or not after registering for notifications that an aurora event is occurring in their location (Kosar et al., 2018). In the aforementioned cases, successful recruitment relies on high levels of sustained engagement where it makes sense to recruit from a pool of known enthusiasts. Thus, several successful citizen science projects have leveraged existing interest groups on social platforms (for example, Aurorasaurus with the Alberta Aurora Chasers Facebook group) or long-existing networks (such as the Citizen Sky Project of the American Association of Variable Star Observers (AAVSO)) (Aaron Price & Lee, 2013).

Other projects may have tasks that require less specialized skill and can rely on brief tutorials plus supporting resources such as discussion boards to train new participants on the required tasks. In this latter case, there is evidence that, due to a lower barrier to entry, more people will participate but may only engage for shorter periods of time (Spiers et al., 2019). While perhaps not ideal, this engagement profile may be perfectly adequate for simple online tasks that aggregate input from multiple independent classifiers. In addition, online projects that share a platform may benefit in the recruitment phase from cross-promotion. Thus, access to substantial "built-in" volunteer audiences is one of the core reasons why platforms such as the Zooniverse, iNaturalist, CosmoQuest, or SciStarter are so important in lowering the barrier for research teams to be able to use the citizen science methodology.

These sorts of differences between project types can account for the wide range in participant numbers: Aurorasaurus registered over 9500 "raw" observations for the period 2015−16 while Backyard Worlds, as of the time of this writing the most popular space science project on Zooniverse, has over 53,000 registered volunteers and collected nearly 600,000 classifications of images in the 2.5 years since launch on February 17, 2017. To be sure, this is not a value statement on the efficacy of Backyard Worlds over Aurorasaurus; the point is to have a clear understanding of the interplay between project goals, task difficulty, and volunteer engagement factors before embarking on running a citizen science project.

Nonetheless, there are many commonalities at the root of project success as defined minimally by retention of volunteers and scientific impact. These parameters of success were studied for a range of Zooniverse projects in Cox et al. (2015), and the results boil down to a key point which can be thought of as the "golden rule" of citizen science: the most successful projects are those where the research team continues to communicate with the volunteer community, keeping participants updated on everything from new data to published results and making team members available for answering questions and discussing science. In fact, these are not new results for the data collection citizen science community where, for example, bird counts have been conducted for over a hundred years using these rules of engagement. In short, it takes time and effort to manage a successful citizen science project, but the reward is at least twofold: (1) high-quality data that can be used in publications and (2) members of the public who are that much more involved in and aware of the process of science.

## Know Your Audience

In designing a citizen science project, it is important to know the characteristics of your volunteer corps, both to understand what is reasonable to expect from them and to learn what may need to be changed about your design. This latter point is crucial for designing projects that could appeal to more diverse audiences, a point to which I will return later in this section. Here, I provide a few details about the Zooniverse volunteers from the 2014 survey cited earlier. Fig. 10.5 shows the 2014 survey respondents' age distribution, when they participate in Zooniverse projects, their

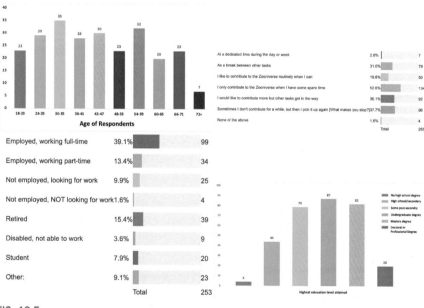

**FIG. 10.5**

Answers to survey questions. Top left: "Please tell us your age." Top right: "When do you participate in the Zooniverse? Please tick all that apply." Bottom left: "Which of the following categories best describes your employment status?" Bottom right: "What is the highest level of school you have completed (combines both United States and United Kingdom and their equivalent systems)?"

*Used with permission from data collected by Homsy, V. (2015). Engagement in the Zooniverse. (Master's thesis, MSc Computer Science). UK: St Cross College, University of Oxford.*

employment status, and their education level. Inasmuch as the respondents are representative of the Zooniverse audience at large, there are a few generalities that can be drawn from these graphs. Zooniverse engages people from a wide range of ages[11] with over 15% retired. In terms of educational attainment, 14% of survey respondents only have a high school degree; 27.5% of respondents had at most a bachelor's degree (or the UK equivalent); and over 32% of respondents had either a master's, doctoral, or professional degree. Compare these figures with the fact that nearly 28% of Americans over 25 have only a high school degree or equivalent, 22% have at most a bachelor's degree, and 13% hold higher degrees,[12] and it becomes evident that the educational attainment of Zooniverse survey respondents is skewed considerably toward greater levels of education. While survey respondents came from 33 separate countries, 35% were from the United States, nearly 30% were from the United Kingdom, and another 10% were from Canada and Australia combined.

Further results from the 2014 survey showed that 38% of the respondents were female, while 60% were male. This is aggregated over all projects in which the 300 respondents participated. A more recent study of 63 Zooniverse projects by Spiers et al. (2019) looks in detail at the gender breakdown for five astronomy and five ecology projects, as shown in Fig. 10.6. It is clear that for the astronomy projects, in aggregate, there is a gender bias toward male participants. Another survey carried out by the CosmoQuest citizen science platform that is predominantly geared toward space science projects obtains similar numbers for age, education, and gender distributions (Gugliucci et al., 2014).

In spite of these gender disparities, I still believe that one of the wonderful consequences of the growth of citizen science over the past decade is the opportunities it presents to really change who participates in "science." As professional scientists have worked hard over the past decade to improve the diversity of their ranks, citizen science can open the doors more widely to encourage diverse audiences to be exposed to the world of science. One possible mechanism described by Brouwer and Hessels, (2019) suggests that targeted recruitment of specific audiences increases both participation of the targeted audiences and the value of citizen science. The 2018 National Academies report *Learning Through Citizen Science: Enhancing Opportunities by Design* (Engineering National Academies, 2018) is a great reference for those interested in further details on how to design citizen science projects for diverse audiences, including how to be mindful of differing cultural perspectives on the fundamental relationship between science and society. While it is imperative that we grow opportunities for underrepresented minorities to participate in science, I can say anecdotally one audience that I know we reach with online projects is people on the autism spectrum or who have physical disabilities which make life very difficult for them. Through my experiences in giving a rather large number of public

---

[11]  Due to legal consent requirements, the survey was not able to solicit responses from anyone under 18 years of age. We know anecdotally that we do have participants as young as 6 years old.

[12]  https://en.wikipedia.org/wiki/Educational_attainment_in_the_United_States.

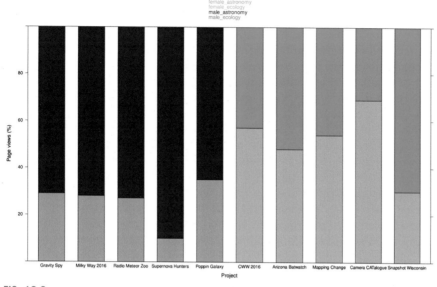

**FIG. 10.6**

The gender breakdown of classification page views for five ecology projects compared to five astronomy projects on the Zooniverse platform. (Originally published in Spiers et al. (2019) and used here under the Creative Commons Attribution 4.0 International (CC BY 4.0) license).

lectures where I have the pleasure of meeting Zooniverse volunteers, I have had several parents come up and touch me deeply with their stories of sons or daughters who have found meaning in their lives through their ability to contribute to science through Zooniverse. We should do everything we can to incorporate accessibility design principles to encourage participation by these audiences in citizen science.

## Learning Through Citizen Science

A remaining question then is: beyond being engaged in the process of research, do citizen scientists increase their understanding of the process of science or scientific content? Is citizen science an effective way to increase scientific literacy? There is a growing corpus of research that shows that participating in citizen science can indeed improve volunteers' scientific literacy and increase their positive attitudes toward science (Aaron Price & Lee, 2013; Brossard et al., 2005; Cox et al., 2015; Crall et al., 2012; Trumbull et al., 2000).

Prather et al. (2013) found that for volunteers that answered questions on a concept inventory designed specifically for Galaxy Zoo, a correlation existed between number of classifications made by volunteers and the correctness of their answers; however, there was no way to determine a causal relationship. Indeed, there are unique difficulties that must be overcome when designing studies to ascertain learning outcomes on citizen science projects. For example, the "free choice," informal nature inherent in most citizen science projects, creates difficulty for designing consistent pre—posttesting conditions when volunteers have a wide range of participation habits. Furthermore, difficulties are likely to arise when asking volunteers to participate in testing strategies that are perhaps more appropriate for formal education learning environments; while rigorous concept inventories with a large number of questions may be useful in assessing learning in fine-grained detail for student cohorts that take a semester-long course, their use in assessing volunteer learning in citizen science projects can interfere with the volunteer experience and may lead to reduced participation in the actual task associated with the science goals of the project. Nonetheless, Masters et al. (2016) show that, by using a limited number of questions and controlling for participation levels and general science knowledge on five separate Zooniverse projects, there is evidence for learning science content related to specific projects. Drawing on additional work from Cox et al. (2015), the results from Masters et al. also indicate that the more the research team engages with the volunteers, the more science content is learned by project participants. While this would seem to be common sense, in describing a model for intentional design for education goals along with science goals, Bonney, Cooper et al. (2009) point out that following this model is costly, requiring at least one full-time staff person to run the project.

Thus, designing for educational goals at the same time as the science goals is a real trade-off that project scientists must make. For example, some projects are very short in duration, and it does not make sense to expend the resources and effort in producing well-designed educational support materials. For research teams that are able to incorporate learning objectives in their project, the 2018 National Academies study on citizen science (Engineering National Academies, 2018) provides an excellent review on these topics. Among many important conclusions and recommendations, the study makes it clear that intentional design for learning is key to successfully attaining educational goals. It may be possible that with their common project design features, platforms such as Zooniverse and CosmoQuest, or project aggregators such as SciStarter, can provide several key elements of educational support materials to considerably reduce the effort in mounting projects that also provide for learning science.

## Even Bigger Data: Why Citizen Science Needs Machine Learning and Why Machine Learning Needs Citizen Science

In 2015, a paper entitled "Rotation-Invariant Convolutional Neural Networks for Galaxy Morphology Prediction" was published describing the winning entry in a machine learning competition on the Kaggle platform (www.kaggle.com) for data prediction contests (Dieleman et al., 2015). The Galaxy Zoo team provided

75,000 images from the Galaxy Zoo 2 (GZ2) project along with the consensus classifications from GZ2 volunteers with the challenge to see if someone, anyone, could write a machine learning algorithm that would reproduce the accuracy of the human labelers. Several attempts with machine algorithms had focused on using attributes such as differences in color to discern the morphologies of galaxies, with blue being correlated with "spiral" or disk-like galaxies and red with "smooth" or elliptical galaxies. These algorithms could reach 80% accuracy levels but limited the science topics to those that were not pursuing, for example, the evolution of blue ellipticals or red spirals. In fact, it was this very topic that inspired the original Galaxy Zoo project. At that time, humans were best at providing robust morphology labels for red spirals or blue ellipticals. With GZ2, the task was to determine even more detailed information about the 250,000 brightest galaxies in the original GZ1 data set so astronomers could study, for example, the evolution of galaxies with bars (Galloway et al., 2015; Masters et al., 2011) or how tightly wound spiral galaxy arms are (Willett et al., 2015). So while Galaxy Zoo volunteers had solved the problem of providing detailed morphological classifications for data on the scales of hundreds of thousands of galaxy images, the Galaxy Zoo team was keenly aware that the data rates of "next-generation" telescope surveys such as the Vera Rubin Observatory would generate billions of galaxy images. A simple back-of-the-envelope calculation makes it clear that even if every single person on the planet participated in classifying galaxies, that would not be enough to tackle the problem. Computers have to be brought into the equation to enable any science that requires accurate information on galaxy morphologies (Martin et al., 2020; Walmsley et al., 2019).

By the time the Galaxy Zoo Kaggle competition was launched in late 2013, the field of machine learning had undergone somewhat of a revolution with the successful application of so-called convolutional neural networks (or CNNs). In short, CNNs use training images that have labels already provided to build up a repertoire, or "model," of common image characteristics that (hopefully) lead to the correct classifications when applied to a new "test" set of images without labels. So there was some expectation from the GZ team that the competition would produce an algorithm that could replicate the human ability to correctly discern detailed differences in the galaxy images. Furthermore, there was hope that by "reverse engineering" the algorithm, the team could learn something fundamental about the physical attributes underlying the morphological differences. So it is not without some real irony that the winning entry did not yield any fundamental insights into astronomy but rather underscored a fundamental rule of machine learning: the more training data you have, the better your algorithm will be. The title of the paper explicitly confesses this: the winner used the fact that galaxies are rotationally invariant on the sky to increase the training data set by a factor of 8, effectively reusing the labels provided by the Galaxy Zoo volunteers.

As shown in Fig. 10.7, it is not just astronomy facing an oncoming deluge of data. As more and more research teams are confronted with the inability to analyze their data sets at scale, they seek help from the citizen scientists on Zooniverse, which at the time of writing has hosted over 220 projects. Many of these same

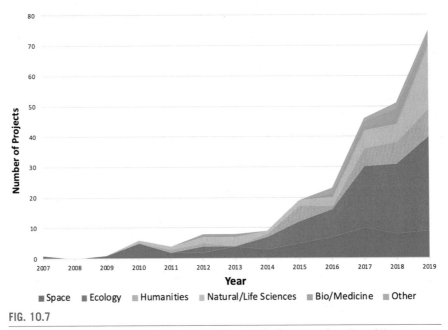

**FIG. 10.7**

The number of projects launched on the Zooniverse platform as a function of time, colored by discipline. Discipline counts are manually coded by the author in part using information provided as tags by the research teams. The data include all projects hosted on the Zooniverse platform up to December 19, 2019.

research teams are also turning to machine learning for help in speeding up their classifications and reducing the involvement of the citizen scientists to analysis of only the most difficult images. In fact, the Galaxy Zoo data as well as many other data sets from Zooniverse are used by countless teams to train machine learning algorithms not just for astrophysics, but for ecology, biomedicine, and even humanities research, as well as basic investigations into improving machine learning (see, for example, Barchi et al. (2020), Willi et al. (2018)). The data deluge is also impacting the data collection citizen science projects, and several have been exploring machine learning as a way to assist in the analysis of the collected data. For example, with Aurorasaurus, there are difficulties in processing the large amounts of Twitter data accurately, and Kosar et al. (2018) suggest that machine learning techniques such as natural language processing algorithms could be useful to speed up the analysis.

Have we solved the problem now for the next-generation data sets? Sort of. Machine algorithms are growing increasingly sophisticated (think self-driving cars) and have recently been shown to do very well compared with human labelers (see, for example, He et al. (2015)). However, the repertoire of image characteristics encoded in a machine learning model may or may not have any relevance to the human perception of the information within the images; for example, a set of training data comprising images of cats and dogs was used successfully to train a model for identifying whether a new image has a supernova candidate in it or not (Wright et al., 2017). This work also showed that human classifiers had a higher purity in their list of supernova candidates, while the machine classifiers produced a more complete list; the combined human–machine classifications produced results better than either could alone. The point is, while arriving at the same classification label as often as possible (by design) on a specific set of images, machine "vision" does not necessarily replicate human vision.

The question remains: what about the "weird and wonderful" classes of objects lurking in a dataset? Machines are only as good as their training data and would have missed, for example, the Green Peas. So while machines can now be used on a large fraction of data where we can train them on "known knowns," we need to find some way of allowing for the serendipitous discovery process. Nominally this would require at least one human to visually inspect each image, but then we are right back to where we started with an intractable problem for the upcoming data sets with billions of images. Current research by the Zooniverse team and others suggests that combining both human and machine classifiers can help solve this problem (see Fortson et al., 2018; Trouille et al., 2019 and references therein). For example, we can use clustering algorithms to look for commonalities in a large number of parameters in a data set and then associate images based on some metric of "likeness." We can then ask volunteers to look at grids of these associations and tell us whether the images really are similar and if they represent a type of object that has not been noticed before (see Wright et al., 2019 for more details). Walmsley et al. (2019) describes another method, currently running on the Galaxy Zoo project, combining humans and machines that can help expedite classification of known objects while holding out the possibility that humans can identify rare or unknown classes of objects. Whatever methods prove workable, it is clear that citizen science needs the help of machine algorithms but, at the same time, machine algorithms need the help of the human volunteers in citizen science projects to deliver not just labels for training but also their superior ability to see the weird and the wonderful in an image and ask questions that can lead to a discovery. The best approach then for the future of citizen science (and, I will state, most areas of knowledge production in the future) is one which combines the strengths of both machines and humans.

## Concluding Remarks and Lessons Learned

When the Zooniverse launched as a platform on December 12, 2009, almost exactly 10 years ago from when I am writing this, we did not know how long it would last. I am constantly amazed and deeply impressed by, and forever thankful for, the dedication and skill of our volunteer corps. I am often asked what made the Zooniverse so successful. The first answer is obvious: we could not have done it without the volunteers. But beyond that, at some level, the answer is we were at the right place at the right time with the right people. More explicitly, those of us guiding the Zooniverse in the earlier years (in particular, Zooniverse founder Chris Lintott) I believe had an intuitive understanding of the underlying principles discussed in this chapter. These principles can be boiled down to the fundamental dyadic relationship between the quality of the research output and the ability to engage and retain our volunteer corps. This led to the willingness to wield ethical precepts as a knife edge in decision-making where rule number one was "never waste volunteer effort." The corollaries to this are requiring research teams to (**1**) have a project that requires the citizen science process (in other words, needing some task performed by humans) and (**2**) for the data produced through volunteer clicks to be used in a publication of some sort. Once a project passed rule number one, the second most important rule came into play: the success of the project depended on the willingness of the research team to engage with the volunteers. It was not enough to develop a project with a great user interface; to retain volunteers, there was an expectation that the research team would keep the volunteers informed of the process of research and be willing to explain the concepts behind the research question the volunteers were helping to answer. These intuitions were later borne out by the Cox et al.'s research described earlier.

A third rule relates to the focused and iterative development process in the early days of Zooniverse (led by Arfon Smith). I strongly believe this is what has led to the longevity of the Zooniverse platform. While some common infrastructure had been established even for the launch of Galaxy Zoo 2 in February 2009, we did not try to design a generalized platform out of the gate. In the first instance, we developed specific tools for specific tasks for research teams that had a clear use case that was also likely to be a common use case across multiple research teams. In this manner, we went from decision tree tasks with Galaxy Zoo to a drawing task with Moon Zoo to a marking task with Solar Storm Watch to an annotation task with Old Weather and survey tasks with Snapshot Safari. We also explored data types including images (Galaxy Zoo), in-browser simulations (Galaxy Zoo: Mergers), video clips (Solar Storm Watch), and audio clips (Whale FM). One of our biggest questions internally was whether volunteers would be willing to try their hand at classifying data presented in actual plots. This was tested on Planet Hunters where images of light curves—varying brightness from a star—were shown and our volunteers came through with flying colors. (Who would not want to try to discover a new world!) Once a tool had been developed for a task or data type, it could be more readily replicated for other research projects. However, the overall process still required dedicated effort from Zooniverse developers and designers. By 2014, there were a large number of research teams who were keen to incorporate a Zooniverse project into their work, and in most cases, these teams could readily use combinations of existing tools for their use cases. It was thus imperative that the research teams be given the tools and opportunity to build their own projects. The lessons learned from building specific tools for specific use cases were thus generalized into the development of the Project Builder, which has been a fantastic success in removing the development bottleneck and enabling the research community to more readily adopt citizen science as a research method (note in Fig. 10.7 the exponential increase of launched projects after July 2015 when the Project Builder was launched). The Project Builder also enables flexible use by research teams. For example, there are many other projects that have used the Project Builder but chose not to be listed on the Zooniverse platform for specific reasons including a desire to keep the project within the research team or data privacy concerns. Furthermore, the Project Builder itself enables Zooniverse to continue iterative design through development of an "experimental" tool tested on a specific project that then can be made available to all Project Builder projects.

Along with provisioning of the Project Builder, the Zooniverse team still takes on custom builds when it makes sense to push the envelope on new task types or strategies. An example of this is the work done for the Gravity Spy project, which implements the concept of "leveling-up" for volunteers to help the LIGO gravitational wave detector identify sources of noise (or "glitches") in the system that masquerade as a potential discovery of a new gravitational wave source (Crowston, 2017; Crowston et al., 2017; Zevin et al., 2017). Providing levels allows volunteers to graduate from the easiest sets of identifying glitch classes to the hardest classes, and with the possibility at the highest level of actually discovering entirely new classes of glitches. Clearly, this concept of leveling up can be generalized to many other projects with use cases where volunteers can participate at varying degrees of difficulty and therefore training. This strategy can also be used to incorporate machines alongside humans—a tactic in fact implemented by the Gravity Spy team.

What are the current concerns for the Zooniverse team? I was recently asked at a meeting how it is possible for Zooniverse to be such a professional platform while being run out of an academic environment by leadership still active in research. It is a testament to the skill and spirit of the Zooniverse development and support team that they are able to build and maintain a platform deserving of such praise and more. The core issue is the in-between world that Zooniverse occupies: the platform's success requires professionalism to develop to the agile standards of what people expect from web products these days, all while being supported through research grants from government agencies and private foundations. There is a common misperception that Zooniverse is a company that somehow makes money from its endeavors. Nothing could be further from the truth. We work hard to bring in research grants that can fund the continued development and improvement of the platform while providing project hosting for our research teams for little or no cost to them. Unfortunately, the grant models of funding agencies do not typically fund maintenance or support of existing infrastructure. Zooniverse is not the only platform that struggles as a technology team with feet planted firmly in academia. As the research world adapts to the crush of massive amounts of data, more and more sophisticated data pipelines and technologies need to be developed and—here is the kicker—maintained at a professional level. This costs real money that is not easily available via research proposals. Thus, there will likely need to be a shift in the relationship between data providers and data consumers driven by new "business models" that incorporate funding for professional software developers and data science staff. It would be a shame if the great strides we have made in engaging the public in science should suffer in that process.

Another challenge that Zooniverse faces, from which more general lessons can be extracted, is the issue of community management—from the perspective of both volunteers and research teams. For any project that wishes to operate at a large scale, it is important that the health of the project is monitored in a data-driven manner (for example, through Google Analytics) and that there is capacity to respond to issues that could have an impact on volunteer engagement or usefulness of data outputs. As the Zooniverse continues to expand hosting even more research teams, we need to strike the right balance in providing tools for the research teams to manage their projects autonomously while at the same time interceding where volunteer engagement is suffering. While the Zooniverse team is currently working on providing better monitoring tools for our research teams, along with further support on processing volunteer-produced data, the role of volunteer moderators cannot be overstated especially in terms of keeping tabs on volunteer engagement. Zooniverse has always relied on wonderful moderators that sometimes appear fortuitously out of the crowd. In some cases, moderators have been with the platform since the Galaxy Zoo days and now moderate on multiple projects, many times helping new projects through their launch phase until a new volunteer appears ready to take on the moderating reins. The moderators help guide newcomers, poke the research teams when their attention may have lapsed, and in general keep the discussions on the "Talk" boards civil and welcoming while encouraging curiosity. The moderator role is thus critical to the success of our projects and one mechanism by which the health of the projects can be monitored. This speaks to the discussion of citizen science being a collective intelligence with individuals participating at levels of expertise that are comfortable for them. That said, research teams cannot make the mistake of relying solely on moderators to keep the volunteers engaged: again, Cox et al. show that the most successful projects are those where members of the research team interact with the volunteers. The moderators can then amplify these interactions but should not be responsible for generating them. The burgeoning area of social science research as applied to understanding engagement in citizen science projects will be very helpful in improving project design, including the role of moderators as well as the type of language and messaging used by projects (see, for example, Segal et al., 2018).

But the biggest challenge is the need to incorporate machine intelligence into citizen science protocols. I am convinced that at some level, the machines will always need humans in the loop, and therefore, online citizen science will not fade away as machines take over. That said, we need to be mindful that our volunteers will certainly have their own opinions about working alongside machines. We must avoid the situation where volunteers are driven away from a project because the machines have classified all the images that are in some way compelling to the volunteers. And yet, we cannot just provide some fraction of images that machines could do on their own just to keep people engaged; this would fall on the wrong side of our ethical knife edge of not wasting volunteer effort. I expect that a technique such as the leveling up strategy used by Gravity Spy will help with this conundrum, where perhaps the more engaging work for novice volunteers is related to a wider range of more compelling images provided at the first level. These images would be those that require some aspect of human validation of the machine classifier, whereas images at the top levels may be those that are too difficult for machines but are then also more intriguing for highly skilled volunteers to take on.

In closing, I look forward to the next decade with real excitement for the ability of citizen science to play a meaningful role in space science research, and to seeing the discoveries made beyond the Green Peas and STEVE. I am confident that citizen science as a methodology will continually evolve by taking into account new research done on both the technical and social challenges of citizen science as we face the onslaught of Big Data and machine learning. If there is one thing I have learned over the past decade, it is that we live in a truly amazing world that has the capacity to promote curiosity and ingenuity in all people sparked by the opportunities to participate in real research through citizen science. Those of us working in the field of citizen science just need to figure out how to keep up and to make sure citizen science as a method stays relevant while our volunteers stay happy.

## Acknowledgments

I would like to thank all of the Zooniverse volunteers for their contributions to all the projects on the platform. I would also like to thank all the members of the Zooniverse development and support team without whom the platform would not exist. I thank my Zooniverse colleagues and coleads, Chris Lintott and Laura Trouille, for their support and always interesting and useful discussions about Zooniverse and the future of citizen science in general. I also owe a huge debt of gratitude to my colleagues and research team members of Zooniverse projects that I am active in with my own research, namely Galaxy Zoo, Muon Hunters, Clump Scout, and Supernova Hunters. It is their excellent modeling of how researchers should engage with volunteers that keeps me trying to follow the advice written here! Finally, I would also like to acknowledge partial support from NSF award #1835530 during preparation of this manuscript.

## References

Aaron Price, C., & Lee, H.-S. (2013). Changes in participants' scientific attitudes and epistemological beliefs during an astronomical citizen science project. *Journal of Research in Science Teaching, 50*(7), 773–801.

Abbott, B., Abbott, R., Adhikari, R., Ajith, P., Allen, B., Allen, G., Amin, R., Anderson, D. P., Anderson, S. B., Anderson, W. G., Arain, M. A., Araya, M., Armandula, H., Armor, P., Aso, Y., Aston, S., Aufmuth, P., Aulbert, C., Babak, S., … Zweizig, J. (January 2009). *Einstein@Home search for periodic gravitational waves in LIGO S4 data* (Vol. 79(2), p. 022001).

Arnold, J., & Clarke, D. J. (2014). What is agency? Perspectives in science education research. *International Journal of Science Education, 36*(5), 735–754.

Banfield, J. K., Andernach, H., Kapińska, A. D., Rudnick, L., Hardcastle, M. J., Cotter, G., Vaughan, S., Jones, T. W., Heywood, I., Wing, J. D., Wong, O. I., Matorny, T., Terentev, I. A., L ópez-Śanchez, A. R., Norris, R. P., Seymour, N., Shabala, S. S., & Willett, K. W. (2016). Radio galaxy zoo: Discovery of a oor cluster through a giant wide-angle tail radio galaxy. *Monthly Notices of the Royal Astronomical Society, 460*(3), 2376–2384.

Barchi, P. H., de Carvalho, R. R., Rosa, R. R., Sautter, R. A., Soares-Santos, M., Marques, B. A. D., Clua, E., Gonc alves, T. S., de Śa-Freitas, C., & Moura, T. C. (2020). Machine and deep Learning applied to galaxy morphology - a comparative study. *Astronomy and Computing, 30*, 100334.

Barnard, L., Scott, C., Owens, M., Lockwood, M., Tucker-Hood, K., Thomas, S., Crothers, S., Davies, J. A., Harrison, R., Lintott, C., Simpson, R., O'Donnell, J., Smith, A. M., Waterson, N., Bamford, S., Romeo, F., Kukula, M., Owens, B., Savani, N., ... Harder, B. (2014). The solar stormwatch CME catalogue: Results from the first space weather citizen science project. *Space Weather, 12*, 657–674.

Bird, T. J., Bates, A. E., Lefcheck, J. S., Hill, N. A., Thomson, R. J., Edgar, G. J., Stuart-Smith, R. D., Wotherspoon, S., Krkosek, M., Stuart-Smith, J. F., Pecl, G. T., Barrett, N., & Frusher, S. (2014). Statistical solutions for error and bias in global citizen science datasets. *Biological Conservation, 173*, 144–154.

Bonney, R., Ballard, H., Jordan, R., McCallie, E., Phillips, T., Shirk, J., & Wilderman, C. C. (2009). *Public participation in scientific research: Defining the field and assessing its potential for informal science education. A CAISE inquiry group report.*

Bonney, R., Cooper, C. B., Dickinson, J., Kelling, S., Phillips, T., Rosenberg, K. V., & Shirk, J. (2009). Citizen science: A developing tool for expanding science knowledge and scientific literacy. *Bioscience, 59*(11), 977–984.

Bowler, B. P., Liu, M. C., Shkolnik, E. L., & Dupuy, T. J. (2013). Planets around low-mass stars. III. A young dusty L dwarf companion at the deuterium-burning limit. *The Astrophysical Journal, 774*, 55.

Boyajian, T. S., Alonso, R., Ammerman, A., Armstrong, D., Asensio Ramos, A., Barkaoui, K., Beatty, T. G., Benkhaldoun, Z., Benni, P., Bentley, R. O., Berdyugin, A., Berdyugina, S., Bergeron, S., Bieryla, A., Blain, M. G., Blanco, A. C., Bodman, E. H. L., Boucher, A., Bradley, M., ... Zsidi, G. (2018). The first post-Kepler brightness dips of KIC 8462852. *Astrophysical Journal Letters, 853*(1), L8.

Boyajian, T. S., LaCourse, D. M., Rappaport, S. A., Fabrycky, D., Fischer, D. A., Gandolfi, D., Kennedy, G. M., Korhonen, H., Liu, M. C., Moor, A., Olah, K., Vida, K., Wyatt, M. C., Best, W. M. J., Brewer, J., Ciesla, F., Csák, B., Deeg, H. J., Dupuy, T. J., ... Szewczyk, A. (April 2016). Planet hunters IX. KIC 8462852-where's the flux? *Monthly Notices of the Royal Astronomical Society, 457*(4), 3988–4004.

Brossard, D., Lewenstein, B., & Bonney, R. (2005). Scientific knowledge and attitude change: The impact of a citizen science project. *International Journal of Science Education, 27*(9), 1099–1121.

Brouwer, S., & Hessels, L. K. (2019). Increasing research impact with citizen science: The influence of recruitment strategies on sample diversity. *Public Understanding of Science, 28*(5), 606–621. PMID: 30995163.

Buxner, S. R., Sharma, M., Hsu, B., Peticolas, L., Nova, M. A. M., & CoBabe-Ammann, E. (August 2012). Barriers, lessons learned, and best practices in engaging scientists in education and public outreach. In J. B. Jensen, J. G. Manning, M. G. Gibbs, & D. Daou (Eds.), *Connecting people to science: A national conference on science education and public outreach, volume 457 of astronomical society of the Pacific conference series* (p. 81).

Cappadonna, J. L., Santos-Lang, C., Duerr, R. E., Virapongse, A., West, S. E., Kyba, C. C. M., Bowser, A., Cooper, C. B., Sforzi, A., Metcalfe, A. N., Harris, E. S., Thiel, M., Haklay, M., Ponciano, L., Roche, J., Ceccaroni, L., Shilling, F. M., Drler, D., Heigl, F., ... Jiang, Q. (2017). Citizen science terminology matters: Exploring key terms. *Citizen Science: Theory and Practice, 2*(1), 1.

Cardamone, C., Schawinski, K., Sarzi, M., Bamford, S. P., Bennert, N., Urry, C. M., Lintott, C., Keel, W. C., Parejko, J., Nichol, R. C., Thomas, D., Andreescu, D., Murray, P., Raddick, M. J., Slosar, A., Szalay, A., & Vandenberg, J. (November 2009). Galaxy zoo green peas: Discovery of a class of compact extremely star-forming galaxies. *Monthly Notices of the Royal Astronomical Society, 399*, 1191−1205.

Coleman, E. A., Ishikawa, S. T., & Gulick, V. C. (March 2014). Clickworkers interactive: Progress on a JPEG2000-streaming annotation interface. In *Lunar and planetary science conference* (p. 2593).

Cooper, S., Khatib, F., Treuille, A., Barbero, J., Lee, J., Beenen, M., Leaver-Fay, A., Baker, D., Popovic, Z., & Players, F. (2010). Predicting protein structures with a multiplayer online game. *Nature, 466*, 756−760.

Cooper, C. B., Shirk, J., & Zuckerberg, B. (2014). The invisible prevalence of citizen science in global research: Migratory birds and climate change. *PloS One, 9*(9), e106508.

Cox, J., Young Oh, E., Simmons, B., Lintott, C., Masters, K., Greenhill, A., Graham, G., & Holmes, K. (2015). Defining and measuring success in online citizen science: A case study of zooniverse projects. *Computing in Science & Engineering, 17*(4), 28−41.

Crall, A., Jordan, R., Holfelder, K., Newman, G., Graham, J., & Waller, D. (April 2012). *The impacts of an invasive species citizen science training program on participant attitudes, behavior, and science literacy* (Vol. 22).

Crowston, K. (2017). Gravity Spy: Humans, machines and the future of citizen science. In *Companion of the 2017 ACM conference on computer supported cooperative work and social computing* (pp. 163−166). ACM.

Crowston, K., Østerlund, C., & Lee, T. K. (2017). Blending machine and human learning processes. In *Proceedings of the 50th Hawaii international conference on system sciences*.

Dang, L., & Russo, P. (2015). How astronomers view education and public outreach: An exploratory study. *Communicating Astronomy with the Public Journal, 18*, 16.

Danielsen, F., Burgess, N. D., & Balmford, A. (2005). Monitoring matters: Examining the potential of locally-based approaches. *Biodiversity & Conservation, 14*(11), 2507−2542.

Davis, C. J., Davies, J. A., St Cyr, O. C., Campbell-Brown, M., Skelt, A., Kaiser, M., Meyer-Vernet, Nicole, Crothers, S., Lintott, C., Smith, A., Bamford, S., & Baeten, E. M. L. (February 2012). *The distribution of interplanetary dust between 0.96 and 1.04 au as inferred from impacts on the STEREO spacecraft observed by the heliospheric imagers* (Vol. 420(2), pp. 1355−1366).

Dieleman, S., Willett, K. W., & Dambre, J. (2015). Rotation-invariant convolutional neural networks for galaxy morphology prediction. *Monthly Notices of the Royal Astronomical Society, 450*, 1441−1459.

Engineering National Academies of Sciences and Medicine. (2018). *Learning through citizen science: Enhancing opportunities by design*. Washington, DC: The National Academies Press.

Fischer, D. A., Schwamb, M. E., Schawinski, K., Lintott, C., Brewer, J., Giguere, M., Lynn, S., Parrish, M., Sartori, T., Simpson, R., Smith, A., Spronck, J., Batalha, N., Rowe, J., Jenkins, J., Bryson, S., Prsa, A., Tenenbaum, P., Crepp, J., ... Zimmermann, V. (2012). Planet hunters: the first two planet candidates identified by the public using the Kepler public archive data. *Monthly Notices of the Royal Astronomical Society, 419*(4), 2900−2911.

Follett, R., & Strezov, V. (2015). An analysis of citizen science based research: Usage and publication patterns. *PloS One, 10*(11), e0143687.

Fortson, L., Masters, K., Nichol, R., Borne, K. D., Edmondson, E. M., Lintott, C., Raddick, J., Schawinski, K., & Wallin, J. (2012). *Galaxy zoo: Morphological classification and citizen science* (pp. 213−236).

Fortson, L., Wright, D., Lintott, C., & Trouille, L. (September 2018). *Optimizing the human-machine partnership with Zooniverse.* arXiv e-prints, page arXiv:1809.09738.

Fraknoi, A. (2005). Steps and missteps toward an emerging profession. *Mercury, 34*(5), 19−25.

Freitag, Amy, Meyer, Ryan, & Whiteman, Liz (2016). Strategies employed by citizen science programs to increase the credibility of their data. *Citizen Science: Theory and Practice, 1.*

Gallardo-Lacourt, B., Liang, J., Nishimura, Y., & Donovan, E. (2018). On the origin of steve: Particle precipitation or ionospheric skyglow? *Geophysical Research Letters, 45*(16), 7968−7973.

Galloway, M. A., Willett, K. W., Fortson, L. F., Cardamone, C. N., Schaw-inski, K., Cheung, E., Lintott, C. J., Masters, K. L., Melvin, T., & Simmons, B. D. (2015). Galaxy zoo: The effect of bar-driven fuelling on the presence of an active galactic nucleus in disc galaxies. *Monthly Notices of the Royal Astronomical Society, 448*(4), 3442−3454.

Garneau, N. L., Nuessle, T. M., Tucker, R. M., Yao, M., Santorico, S. A., Mattes, R. D., & Genetics of Taste Lab Citizen Scientists. (2017). Taste responses to linoleic acid: A crowdsourced population study. *Chemical Senses, 42*(9), 769−775.

Geach, J. E., More, A., Verma, A., Marshall, P. J., Jackson, N., Belles, P.-E., Beswick, R., Baeten, E., Chavez, M., Cornen, C., Cox, B. E., Erben, T., Erickson, N. J., Garrington, S., Harrison, P. A., Harrington, K., Hughes, D. H., Ivison, R. J., Jordan, C., ... Zeballos, M. (2015). The red radio ring: A gravitationally lensed hyper-luminous infrared radio galaxy at z = 2.553 discovered through the citizen science project SPACE WARPS. *Monthly Notices of the Royal Astronomical Society, 452*, 502−510.

Gugliucci, N., Gay, P., & Bracey, G. (July 2014). Citizen science motivations as discovered with cosmo- quest. In J. G. Manning, M. K. Hemenway, J. B. Jensen, & M. G. Gibbs (Eds.), *Ensuring stem literacy: A national conference on STEM education and public outreach, volume 483 of astronomical society of the Pacific conference series* (p. 437).

Hennon, C. C., Knapp, K. R., Schreck, C. J., III, Stevens, S. E., Kossin, J. P., Thorne, P. W., Hennon, P. A., Kruk, M. C., Rennie, J., Gadéa, J.-M., & Striegl, M. (2015). Cyclone center: Can citizen scientists improve tropical cyclone intensity records? *Bulletin of the American Meteorological Society, 96*(4), 591−607.

He, K., Zhang, X., Ren, S., & Sun, J. (February 2015). *Delving deep into rectifiers: Surpassing human-level performance on ImageNet classification.* arXiv e-prints, page arXiv: 1502.01852.

Hoffman, C., Cooper, C. B., Kennedy, E. B., Farooque, M., & Cavalier, D. (2017). Scistarter 2.0: A digital platform to foster and study sustained engagement in citizen science. In *Analyzing the role of citizen science in modern research* (pp. 50−61). IGI Global.

Homsy, V. (2015). *Engagement in the Zooniverse* (Master's thesis, MSc Computer Science). UK: St Cross college, University of Oxford.

Ivezíc, Ž., & The LSST Science Collaboration. (2013). *LSST science requirements document.*

Johnson, L. C., Seth, A. C., Dalcanton, J. J., Wallace, M. L., Simpson, R. J., Lintott, C. J., Kapadia, A., Skillman, E. D., Caldwell, N., Fouesneau, M., & Weisz, D. R. (2015). PHAT stellar cluster survey. II. Andromeda project cluster catalog. *The Astrophysical Journal, 802*(2), 127.

Johnston, A., Hochachka, W. M., Strimas-Mackey, M. E., Ruiz Gutierrez, V., Robinson, O. J., Miller, E. T., Auer, T., Kelling, S. T., & Fink, D. (2019). Best practices for making reliable inferences from citizen science data: Case study using ebird to estimate species distributions. *bioRxiv.*

Keel, W. C., Chojnowski, S. D., Bennert, V. N., Schawinski, K., Lintott, C. J., Stuart Lynn, Pancoast, A., Harris, C., Nierenberg, A. M., Sonnenfeld, A., & Proctor, R. (2012). The galaxy zoo survey for giant AGN-ionized clouds: Ast and resent black hole accretion events. *Monthly Notices of the Royal Astronomical Society, 420*(1), 878–900.

Keel, W. C., Lintott, C. J., Schawinski, K., Bennert, V. N., Thomas, D., Manning, A., Drew Chojnowski, S., van Arkel, H., & Lynn, S. (2012). The history and environment of a faded quasar: Hubble space telescope observations of Hanny's Voorwerp and IC 2497. *The Astronomical Journal, 144*(2), 66.

Kosar, B. C., MacDonald, E. A., Case, N. A., & Heavner, M. (2018). Aurorasaurus database of real-time, crowd-sourced aurora data for space weather research. *Earth and Space Science, 5*(12), 970–980.

Küng, R., Saha, P., Ferreras, I., Baeten, E., Coles, J., Cornen, C., Macmillan, C., Marshall, P., More, A., Oswald, L., Verma, A., & Wilcox, J. K. (2018). Models of gravitational lens candidates from space warps CFHTLS. *Monthly Notices of the Royal Astronomical Society, 474*(3), 3700–3713.

Küng, R., Saha, P., More, A., Baeten, E., Coles, J., Cornen, C., Macmillan, C., Marshall, P., More, S., Odermatt, J., Verma, A., & Wilcox, J. K. (2015). Gravitational lens modelling in a citizen science context. *Monthly Notices of the Royal Astronomical Society, 447*(3), 2170–2180.

Kuchner, M. J., Faherty, J. K., Schneider, A. C., Meisner, A. M., Filippazzo, J. C., Gagće, J., Trouille, L., Silverberg, S. M., Castro, R., Fletcher, B., Mokaev, K., & Stajic, T. (2017). The first Brown dwarf discovered by the backyard worlds: Planet 9 citizen science project. *The Astrophysical Journal Letters, 841*, L19.

Kuchner, M. J., Silverberg, S. M., Bans, A. S., Bhattacharjee, S., Kenyon, S. J., Debes, J. H., Currie, T., Garćia, L., Jung, D., Lintott, C., McElwain, M., Padgett, D. L., Rebull, L. M., Wisniewski, J. P., Nesvold, E., Schawinski, K., Thaller, M. L., Grady, C. A., Biggs, J., ... Disk Detective Collaboration. (2016). Disk detective: Discovery of new circumstellar disk candidates through citizen science. *The Astrophysical Journal, 830*(2), 84.

Lintott, C., Schawinski, K., Bamford, S., Slosar, A., Land, K., Thomas, D., Edmondson, E., Masters, K., Nichol, R. C., Raddick, M. J., Szalay, A., Andreescu, D., Murray, P., & Vandenberg, J. (2011). Galaxy zoo 1: Data release of morphological classifications for nearly 900 000 galaxies. *Monthly Notices of the Royal Astronomical Society, 410,* 166–178.

Lintott, C. J., Schawinski, K., Keel, W., van Arkel, H., Bennert, N., Edmondson, E., Thomas, D., Smith, D. J. B., Herbert, P. D., Jarvis, M. J., Virani, S., Andreescu, D., Bamford, S. P., Land, K., Murray, P., Nichol, R. C., Jordan Raddick, M., Slosar, A.,

Szalay, A., & Vandenberg, J. (2009). Galaxy Zoo: 'Hanny's Voorwerp', a quasar light echo? *Monthly Notices of the Royal Astronomical Society, 399*(1), 129−140.

Lintott, C. J., Schawinski, K., Slosar, A., Land, K., Bamford, S., Thomas, D., Raddick, M. J., Nichol, R. C., Szalay, A., Andreescu, D., Murray, P., & Vandenberg, J. (2008). Galaxy zoo: Morphologies derived from visual inspection of galaxies from the sloan digital sky survey. *Monthly Notices of the Royal Astronomical Society, 389*, 1179−1189.

MacDonald, E. A., Donovan, E., Nishimura, Y., Case, N. A., Megan Gillies, D., Gallardo-Lacourt, B., Archer, W. E., Spanswick, E. L., Bourassa, N., Connors, M., Heavner, M., Jackel, B., Kosar, B., Knudsen, D. J., Ratzlaff, C., & Schofield, I. (2018). New science in plain sight: Citizen scientists lead to the discovery of optical structure in the upper atmosphere. *Science Advances, 4*(3).

Marshall, P. J., Lintott, C. J., & Fletcher, L. N. (2015). Ideas for citizen science in astronomy. *Annual Review of Astronomy and Astrophysics, 53*, 247−278.

Marshall, P. J., Verma, A., More, A., Davis, C. P., More, S., Kapadia, A., Parrish, M., Snyder, C., Wilcox, J., Baeten, E., Macmillan, C., Cornen, C., Baumer, M., Simpson, E., Lintott, C. J., Miller, D., Paget, E., Simpson, R., Smith, A. M., ... Collett, T. E. (2016). Space warps - I. Crowdsourcing the discovery of gravitational lenses. *Monthly Notices of the Royal Astronomical Society, 455*, 1171−1190.

Martin, G., Kaviraj, S., Hocking, A., Read, S. C., & Geach, J. E. (2020). Galaxy morphological classification in deep-wide surveys via unsupervised machine learning. *Monthly Notices of the Royal Astronomical Society, 491*(1), 1408−1426.

Mason Heberling, J., & Isaac, B. L. (2018). Inaturalist as a tool to expand the research value of museum specimens. *Applications in Plant Sciences, 6*(11), e01193.

Masters, K. L., Nichol, R. C., Hoyle, B., Lintott, C., Bamford, S. P., Edmondson, E. M., Fortson, L., Keel, W. C., Schawinski, K., Smith, A. M., & Thomas, D. (2011). Galaxy zoo: bars in disc galaxies. *Monthly Notices of the Royal Astronomical Society, 411*(3), 2026−2034.

Masters, K., Oh, E. Y., Cox, J., Simmons, B., Lintott, C., Graham, G., Greenhill, A., & Holmes, K. (2016). Science learning via participation in online citizen science. *Journal of Science Communication, 15*(03), A07. ArXiv e-prints arXiv:1601.05973.

More, A., Verma, A., Marshall, P. J., More, S., Baeten, E., Wilcox, J., Macmillan, C., Cornen, C., Kapadia, A., Parrish, M., Snyder, C., Davis, C. P., Gavazzi, R., Lintott, C. J., Simpson, R., Miller, D., Smith, A. M., Paget, E., Saha, P., Ku¨ng, R., & Collett, T. E. (2016). Space warps- II. New gravitational lens candidates from the CFHTLS discovered through citizen science. *Monthly Notices of the Royal Astronomical Society, 455*(2), 1191−1210.

Murthy, A. C., Fristoe, T. S., & Burger, J. R. (2016). Homogenizing effects of cities on North American winter bird diversity. *Ecosphere, 7*(1), e01216.

Newman, G. (2014). Citizen cyberscience-new directions and opportunities for human computation. *Human Computation, 1*(2), 103−109.

Newman, G., Graham, J., Crall, A., & Laituri, M. (2011). The art and science of multi-scale citizen science support. *Ecological Informatics, 6*(3−4), 217−227.

Parrish, J. K., Litle, K., Dolliver, J., Hass, T., Burgess, H., Frost, E., Wright, C., & Jones, T. (2017). Defining the baseline and tracking change in seabird populations: The Coastal Observation and Seabird Survey Team (COASST). In J. A. Cigliano, & H. L. Ballard (Eds.), *Citizen science for coastal and marine conservation* (pp. 19−38). New York: Routledge.

Prather, E. E., Cormier, S., Wallace, C. S., Lintott, C., Jordan Raddick, M., & Smith, A. (2013). Measuring the conceptual understandings of citizen scientists participating in zooniverse projects: A first approach. *Astronomy Education Review, 12*(1).

Raddick, M. J., Bracey, G., Gay, P. L., Lintott, C. J., Cardamone, C., Murray, P., Schawinski, K., Szalay, A. S., & Vandenberg, J. (2013). Galaxy zoo: Motivations of citizen scientists. *Astronomy Education Review, 12*(1), 010106.

Raddick, M. J., Bracey, G., Gay, P. L., Lintott, C. J., Murray, P., Schawinski, K., Szalay, A. S., & Vandenberg, J. (2010). Galaxy zoo: Exploring the motivations of citizen science volunteers. *Astronomy Education Review, 9*(1), 010103.

Reges, H. W., Doesken, N., Turner, J., Newman, N., Bergantino, A., & Schwalbe, Z. (2016). Cocorahs: The evolution and accomplishments of a volunteer rain gauge network. *Bulletin of the American Meteorological Society, 97*(10), 1831−1846.

Robbins, S. J., Antonenko, I., Kirchoff, M. R., Chapman, C. R., Fas- sett, C. I., Herrick, R., Singer, K., Zanetti, M., Lehan, C., Huang, D., & Gay, P. L. (May 2014). *The variability of crater identification among expert and community crater analysts* (Vol. 234, pp. 109−131).

Rosendhal, J. D., Sakimoto, P. J., Pertzborn, R. A., & Cooper, L. (December 2004). The NASA office of space science education and public outreach program. In Carolyn Narasimhan, Bernhard Beck- Winchatz, Isabel Hawkins, & C. Runyon (Eds.), *NASA Office of space science education and public outreach conference, volume 319 of astronomical society of the Pacific conference series* (p. 423).

Schawinski, K., Evans, D. A., Virani, S., Megan Urry, C., Keel, W. C., Natarajan, P., Lintott, C. J., Manning, A., Coppi, P., Kaviraj, S., Bamford, S. P., ózsa, G. I. G., Garrett, M., van Arkel, H., Gay, P., & Fortson, L. (November 2010). *The sudden death of the nearest Quasar.* (Vol. 724(1), pp. L30−L33).

Schwamb, M. E., Lintott, C. J., Fischer, D. A., Giguere, M. J., Lynn, S., Smith, A. M., Brewer, J. M., Parrish, M., Schawinski, K., & Simpson, R. J. (August 2012). *Planet hunters: Assessing the Kepler inventory of short-period planets* (Vol. 754(2), p. 129).

Schwamb, M. E., Orosz, J. A., Carter, J. A., Welsh, W. F., Fischer, D. A., Torres, G., Howard, A. W., Crepp, J. R., Keel, W. C., Lintott, C. J., Kaib, N. A., Terrell, D., Gagliano, R., Jek, K. J., Parrish, M., Smith, A. M., Lynn, S., Simpson, R. J., Giguere, M. J., & Schawinski, K. (May 2013). *Planet hunters: A transiting circumbinary planet in a quadruple star system* (Vol. 768(2), p. 127).

Segal, A., Gal, K., Kamar, E., Horvitz, E., & Miller, G. (2018). Optimizing interventions via offline policy evaluation: Studies in citizen science. In *Thirty-second AAAI conference on artificial intelligence*.

Simmons, B. D., Melvin, T., Lintott, C., Masters, K. L., Willett, K. W., Keel, W. C., Smethurst, R. J., Cheung, E., Nichol, R. C., Schawinski, K., Rutkowski, M., Kartaltepe, J. S., Bell, E. F., Casteels, K. R. V., Conselice, C. J., Almaini, O., Ferguson, H. C., Fortson, L., Hartley, W., … Wuyts, S. (2014). Galaxy zoo: CANDELS barred discs and bar fractions. *Monthly Notices of the Royal Astronomical Society, 445*(4), 3466−3474.

Simpson, E., Roberts, S., Psorakis, I., & Smith, A. (2013). *Dynamic bayesian combination of multiple imperfect classifiers* (pp. 1−35). Berlin, Heidelberg: Springer Berlin Heidelberg.

Smith, D. A., Peticolas, L., Schwerin, T., Shipp, S., & Manning, J. G. (July 2014). Science and science education go hand-in-hand: The impact of the NASA science mission directorate educa- tion and public outreach program. In J. G. Manning, M. K. Hemenway, J. B. Jensen, & M. G. Gibbs (Eds.), *Ensuring stem literacy: A national conference on*

*STEM education and public outreach, volume 483 of astronomical society of the Pacific conference series* (p. 9).

Straub, M. C. P. (2016). Giving citizen scientists a chance: A study of volunteer-led scientific discovery. *Citizen Science: Theory and Practice, 1*(1), 5.

Sullivan, B. L., Aycrigg, J. L., Barry, J. H., Bonney, R. E., Bruns, N., Cooper, C. B., Damoulas, T., Dhondt, A. A., Dietterich, T., Farnsworth, A., Fink, D., Fitzpatrick, J. W., Fredericks, T., Gerbracht, J., Gomes, C., Hochachka, W. M., Iliff, M. J., Lagoze, C., La Sorte, F. A., … Kelling, S. (2014). The ebird enterprise: An integrated approach to develop- ment and application of citizen science. *Biological Conservation, 169*, 31–40.

Sullivan, B. L., Wood, C. L., Iliff, M. J., Bonney, R. E., Fink, D., & Kelling, S. (2009). ebird: A citizen-based bird observation network in the biological sciences. *Biological Conservation, 142*(10), 2282–2292.

Spiers, H., Swanson, A., Fortson, L., Simmons, B. D., Trouille, L., Blickhan, S., & Lintott, C. (2019). Everyone counts? Design considerations in online citizen science. *Journal of Science Communication, 18*(1), A04.

Swanson, A., Kosmala, M., Lintott, C., Simpson, R., Smith, A., & Packer, C. (2015). Snapshot serengeti, high-frequency annotated camera trap images of 40 mammalian species in an African savanna. *Scientific Data, 2*(150026).

Theobald, E. J., Ettinger, A. K., Burgess, H. K., DeBey, L. B., Schmidt, N. R., Froehlich, H. E., Wagner, C., HilleRisLambers, J., Tewksbury, J., Harsch, M. A., & Parrish, J. K. (2015). Global change and local solutions: Tapping the unrealized potential of citizen science for biodiversity re- search. *Biological Conservation, 181*, 236–244.

Thiry, H., Laursen, S. L., & Hunter, A.-B. (2008). Professional development needs and outcomes for education-engaged scientists: A research-based framework. *Journal of Geoscience Education, 56*(3), 235–246.

Trouille, L., Lintott, C. J., & Fortson, L. F. (2019). Citizen science frontiers: Efficiency, engagement, and serendipitous discovery with human–machine systems. *Proceedings of the National Academy of Sciences, 116*(6), 1902–1909.

Trumbull, D. J., Bonney, R., Bascom, D., & Cabral, A. (2000). Thinking scientifically during participation in a citizen science project. *Science Education, 84*(2), 265–275.

Walmsley, M., Smith, L., Lintott, C., Gal, Y., Bamford, S., Dickinson, H., Fortson, L., Kruk, S., Masters, K., & Scarlata, C. (2019). Galaxy zoo: Probabilistic morphology through bayesian CNNs and active learning. *Monthly Notices of the Royal Astronomical Society, 491*. arXiv e-prints, page arXiv:1905.07424.

Westphal, A. J., Anderson, D., Butterworth, A. L., Frank, D. R., Lettieri, R., Marchant, W., Von Korff, J., Zevin, D., Ardizzone, A., Campanile, A., Capraro, M., Courtney, K., Criswell, M. N., III, Crumpler, D., Cwik, R., Jacob Gray, F., Hudson, B., Imada, G., Karr, J., … Zolensky, M. E. (2014). Stardust interstellar preliminary examination i: Identification of tracks in aerogel. *Meteoritics & Planetary Sciences, 49*(9), 1509–1521.

Westphal, A. J., Stroud, R. M., Bechtel, H. A., Brenker, F. E., But- terworth, A. L., Flynn, G. J., Frank, D. R., Gainsforth, Z., Hillier, J. K., Postberg, F., Simionovici, A. S., Sterken, V. J., Nittler, L. R., Allen, C., Andn, D., Ansari, A., Bajt, S., Bastien, R. K., Bassim, N., … Zolensky. (2014). Evidence for interstellar origin of seven dust particles collected by the stardust spacecraft. *Science, 345*(6198), 786–791.

Wiggins, A., Newman, G., Stevenson, R. D., & Crowston, K. (December 2011). Mechanisms for data quality and validation in citizen science. In *2011 IEEE seventh international conference on e-science workshops* (pp. 14–19).

Willett, K. W., Galloway, M. A., Bamford, S. P., Lintott, C. J., Masters, K. L., Scarlata, C., Simmons, B. D., Beck, M., Cardamone, C. N., Cheung, E., Edmondson, E. M., Fortson, L. F., Griffith, R. L., H äußler, B., Han, A., Hart, R., Melvin, T., Parrish, M., Schawinski, K., Smethurst, R. J., & Smith, A. M. (2017). Galaxy zoo: Morphological classifications for 120 000 galaxies in HST legacy imaging. *Monthly Notices of the Royal Astronomical Society, 464*, 4176–4203.

Willett, K. W., Schawinski, K., Simmons, B. D., Masters, K. L., Skibba, R. A., Kaviraj, S., Melvin, T., Ivy Wong, O., Nichol, R. C., Cheung, E., Lintott, C. J., & Fortson, L. (2015). Galaxy zoo: The dependence of the star formation-stellar mass relation on spiral disc morphology. *Monthly Notices of the Royal Astronomical Society, 449*(1), 820–827.

Willi, M., Pitman, R. T., Cardoso, A. W., Locke, C., Swanson, A., Boyer, A., Veldthuis, M., & Fortson, L. (2018). Identifying animal species in camera trap images using deep learning and citizen science. *Methods in Ecology and Evolution, 10*(1), 80–91.

Wright, D. E., Fortson, L., Lintott, C., Laraia, M., & Walmsley, M. (2019). Help me to help you: Machine augmented citizen science. *Transactions on Social Computing, 2*(3), 11: 1–11:20.

Wright, D. E., Lintott, C. J., Smartt, S. J., Smith, K. W., Fortson, L., Trouille, L., Allen, C. R., Beck, M., Bouslog, M. C., Boyer, A., Chambers, K. C., Flewelling, H., Granger, W., Magnier, E. A., McMaster, A., Miller, G. R. M., O'Donnell, J. E., Simmons, B., Spiers, H., … Young, D. R. (2017). A transient search using combined human and machine classifications. *Monthly Notices of the Royal Astronomical Society, 472*(2), 1315–1323.

Zevin, M., Coughlin, S., Bahaadini, S., Besler, E., Rohani, N., Allen, S., Cabero, M., Crowston, K., Katsaggelos, A. K., Larson, S. L., Lee, T. K., Lintott, C., Littenberg, T. B., Lundgren, A., Østerlund, C., Smith, J. R., Trouille, L., & Kalogera, V. (2017). Gravity spy: Integrating advanced ligo detector characterization, machine learning, and citizen science. *Classical and Quantum Gravity, 34*(6), 064003.

# From Management to Engagement: How South Africa's Square Kilometer Array Project Transformed Its Interactions With Stakeholder Groups

# 11

Anton Binneman, Corné Davis

## Introduction

The success of major infrastructural projects often depends on effective stakeholder engagement and communication. The ever-changing project environment and stakeholders necessitate the continuous evaluation of the engagement methods to ultimately minimize risk and maximize success. This is especially true for a multinational project, like the SKA. The square kilometer array (SKA) project is an international effort to build the world's largest radio telescope. In 2010, South Africa's arid Karoo region was selected to host the core of the midfrequency antennas of the SKA (others will extend over the African continent), and Australia's Murchison Shire was selected to host the low-frequency antennas. These regions were chosen as cohosting locations for many scientific and technical reasons, ranging from the atmospherics above the sites to the radio quietness in these remote areas, as can be seen in Fig. 11.1A and B.

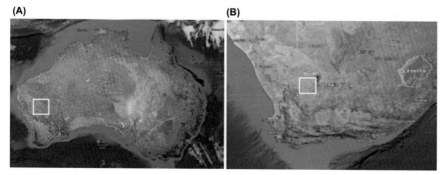

**FIG. 11.1**

(A) Australia and (B) South Africa locations of the telescope.

Space Science and Public Engagement. https://doi.org/10.1016/B978-0-12-817390-9.00006-3

As one of the largest scientific endeavors in history, the SKA project involves a wealth of the world's leading scientists, engineers, and policymakers. However, the establishment of the SKA project holds both positive and negative consequences that affect these and other stakeholders differently. Stakeholders in this instance refer to any person, organization, body or forum that can affect or be affected by the project, ranging from the scientific sector at the one end to local communities at the other. This chapter reflects on how stakeholder communication has evolved since the inception of the project and sheds light on the complexities and challenges of science communication and stakeholder communication. The discussions on stakeholder communication that developed through different phases, as identified, demonstrate how the shift from the *management of stakeholders* to *engagement with stakeholders* occurred and the subsequent implications thereof.

The project is rich in multinational and local stakeholders that have significant power. These groups vary in their attitudes toward the project; some are invested in the project, and in most cases, they tend to adopt an unsolicited attitude of support. But negative attitudes toward the project do exist. This holds true for some local groups that will be discussed later in the chapter. As a result, SKA's managing entity in South Africa, the National Research Foundation/South African Radio Astronomy Observatory (NRF|SARAO), has to interact with its stakeholders constantly and strategically.

*Stakeholder management* tends to be a top-down approach, based on the assumption that stakeholders can be controlled, to some extent, by the organization managing a project and for the purposes of the organization. This is often done through one-way communication, where stakeholders are informed rather than engaged. Instead, *stakeholder engagement* is multidirectional. The discussion that follows shows how classic stakeholder control through communication evolved into stakeholder engagement through the different communication strategies that were employed for each stakeholder group. This story reflects the experience of and path forged by one of us, Dr. Anton Binneman, who serves as stakeholder manager for the project at SARAO, with the academic research support of the other, Dr. Corné Davis. We hope to show through our experience of engaging with stakeholders on different platforms and through different methods, all while constantly reflecting on academic best practices, how to include stakeholders in a space science project in a sustainable manner.

## The Square Kilometer Array in South Africa

South Africa has a rich history of astronomy and has frequently collaborated with international astronomers over the past 100 years. These collaborations were mostly in optical astronomy, although there have been collaborations in radio astronomy as well. Built in 1961 by NASA as the Deep Space Station 51, the Hartebeesthoek Radio Astronomy Observatory (HartRAO) is a 26-m-diameter antenna that was used to retrieve data from and send commands to many unmanned US space probes beyond

Earth's orbit. These included the Ranger, Surveyor and Lunar Orbiter spacecraft that either landed on the Moon or mapped it from orbit; the Mariner missions that explored the planets Venus and Mars; and the Pioneers that measured the Sun's winds.

The station was handed over to the South African Council for Scientific and Industrial Research in 1975 and was then converted into a radio astronomy observatory. In 1988, the observatory became a national facility, operated by the Foundation for Research Development (FRD). The FRD was subsequently restructured in 1999 and reestablished as the National Research Foundation (NRF), a public benefit organization and entity of the South African Department of Arts, Culture, Science, and Technology. The main purpose of the NRF was and still is to capacitate research and research excellence in South Africa, including in the fields of astronomy and radio astronomy.

Concurrently, the concept behind the SKA emerged from the multinational Large Telescope Working Group in an effort to develop specifications for a next-generation radio telescope. In 1997, eight institutions from six countries including Australia, Canada, China, India, the Netherlands, and the United States of America signed a Memorandum of Agreement (MOA) to cooperate in a technology study program that would produce a very large radio telescope in the future. This was superseded by an MOA in 2005 to collaborate in the development of the SKA, increasing the number of stakeholders from 8 institutions to 21 from various countries. A multinational partnership was formed. As a result, the International SKA Project Office was established in Manchester, United Kingdom, and a call for proposals to host the SKA radio telescope was published in 2009.

In response, South Africa's HartRAO embarked on an ambitious radio astronomy program to participate in the bidding process to host the SKA radio telescope, constructing KAT-7, the 7-dish precursor of the 64-dish MeerKAT telescope array in the Karoo region. KAT-7 acted as a technology demonstrator to convince international astronomy bodies that South Africa had the capacity to build instruments of this nature. In 2010, Australia and South Africa were identified as finalists and subsequently named as cohosts for the instrument, and SKA South Africa came into existence.

This proposed radio telescope would have the ideal location. In 2007, South Africa's parliament passed the Astronomy Geographic Advantage Act, which declared the Northern Cape province in South Africa an "astronomy advantage area," protecting it from future radio interference. The Karoo region is considered the perfect location for the SKA project because it is protected, remote, and sparsely populated, with a very dry climate. The area has minimal radio frequency interference from human-made sources, such as cellular phones and broadcast transmitters, which is crucial for radio astronomy. Radio interference would "blind" the telescope and essentially render it ineffective.

As a result, an area of 12.5-million hectares around the proposed core of the SKA is protected as a radio astronomy reserve, with regulations controlling the generation and transmission of interfering radio signals in and around the area. This protection was strengthened when additional regulations were passed as part of the Astronomy Geographic Advantage Act in 2018 to ensure that the site remains optimal for radio astronomy specifically. With the area protected, the NRF procured 120,000 hectares of land surrounding the core area to optimize operations. It was subsequently decided that the procured land and infrastructure would be managed by the NRF in collaboration with the South African National Parks Authority.

In 2017, the then South African Minister of Science and Technology announced that SKA SA and HartRAO would be combined to form the NRF's national facility, the South African Radio Astronomy Observatory (NRF|SARAO). Thus, all radio astronomy was consolidated under NRF|SARAO with an amended mandate. At present, NRF|SARAO spearheads SKA activities in engineering, science, and construction in South Africa. In addition, NRF|SARAO incorporates radio astronomy instruments and programs such as the MeerKAT and KAT-7 telescopes in the Karoo, HartRAO in Gauteng, the African Very Long Baseline Interferometry program situated in nine African countries, as well as the human capital development and commercialization endeavors that support these projects.

Eleven countries are currently members of the organization and provide centralized leadership for the SKA, including Australia, Canada, China, Germany, India (associate member), Italy, New Zealand, South Africa, Sweden, the Netherlands, and the United Kingdom. While the 11 member countries are the cornerstone of the SKA, around 100 organizations from approximately 20 countries are now participating in its design and development.

Eventually, the SKA will use hundreds of dishes and hundreds of thousands of low-frequency aperture array telescopes. Instead of the telescopes being clustered in the central core regions, they will be positioned in spiral-arm configurations over a vast distance extending from the core. This positioning will create a long-baseline interferometer array, where all the instruments will act as a single dish to expand the mapping range of the observable universe that could potentially answer profound questions in astrophysics and cosmology.

With this consideration, the project requires the involvement of world-leading scientists and engineers who can design and develop a system with supercomputers that are faster than any in existence at this time, as well as network technology that will generate more data traffic than the entire global system of interconnected computer networks or the Internet. The enormous scale of the SKA project shows the immense progress in both engineering and research and highlights the development toward building and delivering a radio telescope that is likely to produce an exponential and transformational increase in scientific capacity. This is globalization in the field of astronomy, moving toward an informed and interconnected future.

## The Need for Stakeholder Engagement and Communication

The SKA has numerous different stakeholders. This chapter will mention some of these, but the focus of this chapter will be on local stakeholders. Considering the magnitude and significance of the SKA project, effective communication with all local stakeholders, in particular, is considered critical, and sketching the stakeholder landscape is key to providing readers with some insight. As shown in Fig. 11.2, the Northern Cape is the biggest province in South Africa and has the smallest population, consisting of 13% of the country's landmass area and inhabited by only 3% of the country's population, resulting in a population density of three people per square kilometer.

The regional population consists of a multiracial ethnic group considered to be descendants of the indigenous San people. Economic activities include tourism, mining, agriculture, and fisheries. The province is mostly arid, as seen in Fig. 11.3, with two river systems flowing through the province to support agriculture. With high unemployment rates, approximately 35% of the communities are dependent on state grants and consistently face socioeconomic problems such as poverty, domestic violence, and fetal alcohol syndrome, to mention a few. It is a harsh environment.

Knowing that the SKA would have an impact on the local communities, likely stakeholders were identified and recorded in a database at a very early stage of the project. All parties from the communities in the astronomy advantage area that would likely be affected were contacted and invited to public information sessions. The purpose of the information sessions was to ensure that communities and stakeholders at all levels were informed of the potential impacts of the project; the

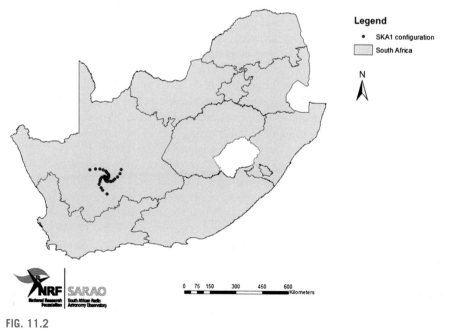

**FIG. 11.2**

SKA's location in the Northern Cape Province in South Africa. *SKA.*

FIG. 11.3

The landscape where SKA will be hosted, with MeerKAT in the distance. *SKA.*

potential mitigation of these impacts; potential opportunities linked to the project; science applied by the project; and construction, science, and community development activities. To contextualize the complexity of the information that had to be shared with the stakeholders, the scope of the impact needs to be explicated. The possible negative effects on the communities include the potential impact on mobile telecommunications, the limiting of radiofrequency interference close to the core of the instrument, the impact on landowners from whom land would be procured, the economic impact on the communities as a result of the procurement of 120,000 hectares of land, the environmental impact related to the construction of the instrument, and the possible growth in the local population. On the other hand, the project offered potential employment opportunities for local communities, economic growth through initiatives to mitigate the negative effects and through foreign direct investment, growth in property value, investment in the development of human capital, entrepreneurship opportunities, and investment in local communities.

Although every effort was made to identify and outline potential impacts, given the diversity of stakeholders that could affect this project and could be affected by it, stakeholders could not simply be managed with disconnected, uniform, one-way communication. Instead, developing an appropriate strategy required the categorization of the different stakeholders and the selection of the most suitable platforms and methods for the various communications and engagements (Table 11.1).

Both the challenges and opportunities that emerged as the project developed demonstrated why there has been a shift in stakeholder theorizing over recent years (Cornelissen, 2014). It became obvious that a pure management approach would not suffice and that the various stakeholders needed to be engaged in dialogue throughout the project. A key shift has been from predominantly organization-focused to issue-focused stakeholder engagement (Roloff, 2008). These shifts have recently been reconceptualized succinctly from the perspective of the emerging

**Table 11.1** Stakeholders Identified in the SKA South Africa Project.

| Group 1 Internal | Group 2 External | Group 3 Community |
|---|---|---|
| SKA Rosebank office/ Johannesburg | Media | International (all individuals interested in the science) |
| SKA Pinelands/ Cape Town office | Researchers | National (all South African citizens) |
| SKA Klerefontein/ Karoo support base | Politicians | Provincial government (Northern Cape) |
| SKA scientists | Department of Science and Technology | Local government (Hantam, Kareeberg and Hoogland) |
| SKA Observatory/ Manchester | Department of Basic Education | NGOs and faith-based organisations |
| SKA contractors | Cooperative Governance and Traditional Affairs (COGTA) and rural development | Educational institutions (Sol Plaatje University, schools and FET colleges) |
| | Universities | Community members |
| | Environmentalists | Farmers |
| | Agri South Africa and farmers' unions | Farmworkers |
| | San Council/other indigenous bodies | Science tourists/tourists |

paradigm of strategic communication (Overton-de Klerk & Verwey, 2013), with emphasis on key notions such as dialogue, inclusivity, collaboration, accountability, and influence. Sloan (2009) agrees that stakeholders who are not directly part of the organization, such as local communities, nongovernmental organizations, and activist groups, have also gained increased attention. Similarly, indigenous minority groups have been featuring more prominently in stakeholder discourses (Corntassel, 2008). Considering this, stakeholders were identified for the purpose of developing stakeholder engagement strategies for the SKA (Table 11.2).

**Table 11.2** Municipalities and Towns Directly Affected by the SKA Project in South Africa.

| District Municipality | Local Municipalities | Towns Included | Towns Affected |
|---|---|---|---|
| Namaqua | Karoo Hoogland Hantam | Williston, Fraserburg, Sutherland, Calvinia, Brandvlei, Swartkop, Loeriesfontein, Niewoudtville, Middelpos | Williston, Fraserburg, Brandvlei and Swartkop |
| Pixley Ka Seme | Kareeberg | Carnarvon, Vanwyksvlei, Vosburg | Carnarvon and Vanwyksvlei |

This chapter explains how the specific communication and engagement strategies for the different local groups identified evolved. Early on, the local public had the expectation that all social and economic issues experienced in their communities would be addressed by the project. Although socioeconomic development of the area is important, this could not be the primary objective of the project, which is to construct a telescope. Nonetheless, within the mandate of the project, a strategy for community engagement had to be developed that addressed the main concerns expressed by the local public. The objectives identified by NRF|SARAO to guide stakeholder communication with different groups in local communities include the following:

- **Youth development.** Human capacity for the instrument needs to be developed. Part of South Africa's contribution to the SKA is to maintain the instrument; some of these skills could be developed and sourced from local communities.
- **Education.** Interest in mathematics and science at school level has to be stimulated and proficiency improved. Bursaries (scholarships) and development opportunities are provided through the SKA project to youth identified in local communities as well as the rest of South Africa.
- **Economic development.** Considering that the local communities are mostly dependent on government grants, and the land acquisitions affected the agricultural industry to a degree, it is crucial for the SKA project to develop local economies and improve their self-sufficiency.
- **Connectivity.** The SKA project limits telecommunication near the core. All affected homesteads have to be equipped with supplementary communications systems. Although a fixed-line and mobile solution was rolled out, these systems are revised on a biannual basis. The focus is on the farming communities because the towns will not be affected.
- **Community development.** One of the decisions that was made by the South African government was to ensure that local communities are improved through the SKA project. Therefore, socioeconomic investment is made in the communities in the astronomy advantage area. This is mainly done by way of grants focused on community upliftment programs such as feeding schemes and soup kitchens and addressing issues such as fetal alcohol syndrome.

Five guiding principles were used to implement these objectives and engage with stakeholders to ensure optimal collaboration, as shown in Fig. 11.4. These principles were adapted from international best practices and demonstrate the project's commitment to ensuring that the communities are fully informed (Table 11.3).

Based on the objectives for the local communities and the guiding principles that were established, a strategy was developed that committed NRF|SARAO and the SKA to stakeholder engagement and integrated stakeholder activities as a core part of the greater organization. Contextually, the aim is to effectively communicate and increase public access to SKA-related information that is accurate, responsible, reliable, and end-user focused. This includes updates on construction, scientific findings, and progress made in terms of community development and opportunities for

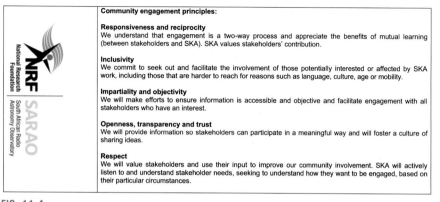

FIG. 11.4

SKA guiding principles. *SKA.*

**Table 11.3** Positive and Negative Impacts of the SKA Project.

| Positive Impacts | Negative Impacts |
| --- | --- |
| Large capital investment in local businesses. (R320 million spent in the Karoo) | Radio frequency limitations, especially in the core (i.e., there will be an impact on telecommunications) |
| 100 people from local communities permanently employed | Visual pollution. |
| Foreign direct investment through projects like the SKA | 130,000 hectares of agricultural land removed from the economy |
| 7500 contract employment positions created and more expected for SKA phase one | Loss of R13 million in potential revenue |
| Hospitality and tourism expanded ten-fold | |
| Bursaries and schools' programs | |
| Science engagement and human capital development in most schools. | |
| 130,000-hectare special reserve declared | |
| Alternative communication provided to affected farmers | |
| Community development initiatives | |
| Greater involvement from other government departments | |
| Increased police capacity due to it being a national key point | |
| Improved communication systems for emergency services | |

local people. The methods for engagement and communication are aligned to the unique requirements of each stakeholder group to ensure a customized and coherent approach to engagement for all NRF|SARAO stakeholders. The following subsections describe the engagement methods used for different stakeholder groups.

## Addressing Political Concerns: Northern Cape, Regional and Local Governments

One of NRF|SARAO's key stakeholders is the government. The South African government is divided into four tiers: national, provincial, regional, and local. These tiers have legislative and executive authority in their own spheres and are defined in the Constitution as "distinctive, interdependent, and interrelated." There are 278 municipalities in South Africa, comprising 8 metropolitan, 44 district, and 226 local municipalities. These municipalities are mandated to develop local economies and to provide infrastructure and services. Therefore, it is important to ensure that engagements lead to political and administrative buy-in.

The national and provincial tiers have been supportive of the SKA project. However, the project had to secure support to enable collaboration with the sectoral departments within the Northern Cape Provincial Government specifically. A large science infrastructure project would have numerous government stakeholders with varied interests and influence, and resistance from any department would have a negative impact on the project deliverables. For example, considering the remote location of the project, road infrastructure would require upgrading and regular maintenance to enable the safe transfer of multimillion-dollar resources and human capital; health facilities and emergency response would need to be improved to ensure responsive medical services for the multinational crew; safety and security would have to be integrated to protect a national key point; spatial planning and rural development would have to be considered; education and training of local people to adequately address the project's needs would have to be looked at; and tourism and local economic development would need to be updated. Thus, all government departments have a role to play, and appropriate platforms for engagement had to be put in place to ensure collaborative efforts for the overall success of the project.

The project instituted three different platforms to ensure that this engagement was done properly. The first was through the national government and the positioning of the MeerKAT and SKA project as a national priority. Since the project forms a crucial part of the South African Government's National Development Plan, political buy-in was ensured on all levels of government, including parliament, opposition parties, and the entire cabinet. The South African Government has identified the construction of the SKA as a strategic infrastructure project overseen by the Presidential Infrastructure Coordinating Committee. The Government foresaw that this would lead to new innovations in manufacturing and construction. The project was positioned from early on as having the potential to transform South Africa's economy through human capital development, innovation, value addition, industrialization, and entrepreneurship. Due to their interest and influence, national government stakeholders are kept updated on the developments of the project and are invited to conduct site visits on a regular basis.

The second platform developed is direct contact with the premier's office through regular updates by the SKA stakeholder manager and the Department of Science and Technology. The project briefed the Northern Cape Government on an individual and parliamentary level. The third platform is a provincial stakeholder forum that was established in collaboration with the Department of Science and Technology. This forum is known as the Northern Cape Working Group. The purpose of this forum is to coordinate activities and manage risks to the project as well as to maintain relationships with the provincial sector departments. The relevant sectors are all represented and meet on a monthly basis. At these meetings, the NRF|SARAO stakeholder manager communicates possible impediments to the project. The resultant outcomes are then communicated to the provincial, regional, and local governments to ultimately stimulate action or support while observing political protocol. An appointed champion from the Department of Economic Development and Tourism currently coordinates the meetings and ensures participation by the stakeholders on behalf of the Northern Cape Government and NRF|SARAO.

The SKA project falls within the boundaries of two district municipalities and three local municipalities, with the towns of Williston and Fraserburg in Karoo Hoogland municipality, Carnarvon and Vanwyksvlei in Kareeberg, and Brandvlei and Zwartkop in Hantam, as depicted in Fig. 11.5. This figure also indicates the core and the spiral arms of the SKA midfrequency antennas. An important project strategy is to engage with local and district municipalities and to involve them on

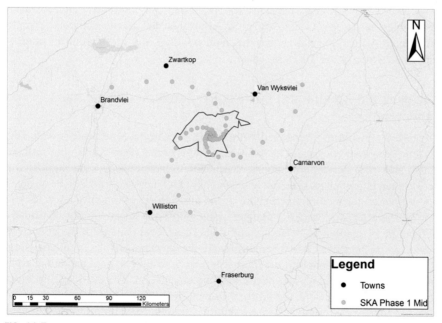

**FIG. 11.5**

Localities directly affected by the SKA project in South Africa. *SKA.*

all relevant engagement platforms. The reason for this is that these municipalities are managed by the elected representatives of the local communities in specific geographical areas.

In many regards, the relationships with local municipalities are more intensive than those with the other tiers of government. The Kareeberg and the Karoo Hoogland municipalities are engaged on a weekly basis, since the 120,000 hectares of land that was purchased from farmers is situated within the domains of these two municipalities. The municipalities also form part of the Northern Cape Working Group; however, local government has full autonomy to govern all community affairs, provided that it does not contravene provincial or national legislation. Resultantly, this stakeholder group has a significant influence on the management of the environment and communities. Therefore, the SKA project stakeholder manager is responsible for regular, direct, and active stakeholder engagement for the purpose of maintaining relationships. This proactive approach with the local government was written into the engagement strategies of the SKA, as well as the SKA Integrated Environmental Management Plan. The intended outcome is to ensure that all groups are included in communication generated by the project and to ensure transparency in terms of the project's impact on the communities.

At present, the relationship is mutually beneficial. The municipalities and the project collaborate in the municipalities' plans for local development needs and land use aimed at promoting sustainable, functional, and integrated human settlements; maximizing resource efficiency; and enhancing regional identity and character. The project provides input to these processes and contributes both financial and human resources. Conclusively, SKA ensures that all its activities comply with South African legislation and that municipalities are informed of all project-related activities in the different areas.

## Engaging the Agricultural Community: Farmers' Unions and Landowners

Organized agriculture is engaged with the SKA project on multiple levels. This stakeholder group consists of national Agri South Africa (Agri SA), Agri Northern Cape, and local agricultural unions. NRF|SARAO acknowledges the agricultural sector as an important stakeholder and contends that landowners need to be engaged and supported for the optimal welfare of the agricultural community, especially in the Northern Cape. Agri SA promotes profitability and sustainability of primary agriculture by maintaining stability and supporting development for the collective, whereas Agri Northern Cape is responsible for the development of agriculture within the province. These nonprofit organizations also contribute to national and international agricultural policies and oversee its implementation. The third level includes local farming unions. Here, the farmers form clusters to optimize agricultural activities and address environmental sustainability within the local farming communities.

Although local agriculture makes a relatively low direct contribution to South Africa's gross domestic product, it is one of the largest contributors to employment and accounts for a significant portion of total exports. The development of the SKA requires large portions of land, and the 120,000 hectares of land NRF|SARAO procured was previously used for sheep farming. The implication of this procurement was that agriculture would be affected. NRF|SARAO shares the global concern around food security and therefore acknowledges the importance of continuous engagement with this sector.

Organized agriculture was involved on different levels to set the scene for engagement with the local agricultural community. A Memorandum of Understanding (MOU) that establishes a partnership and outlines common lines of action and engagement was signed with Agri SA. The purpose of the MOU was to agree on areas for possible cooperation that would be mutually beneficial and in the interest of local communities, including the farming community and the country, and ways to jointly establish sustainability initiatives in the affected areas like the rendering of waste in the slaughtering of sheep. This cooperative effort focuses on activities such as exploring the accommodation of ongoing farming on the acquired land, aligned with applicable legislation and policies; identifying economic opportunities in the affected areas for all communities, including farming communities; undertaking corporate social investment projects in the affected areas by establishing a corporate vehicle jointly administered by NRF|SARAO and Agri SA; solving environmental challenges such as predator and invasive flora control; identifying socioeconomic or environmental areas requiring research; and establishing effective joint communication protocols while ensuring continued access to effective alternative telecommunication services for the affected communities.

As with other stakeholders, Agri SA and Agri Northern Cape are consulted regularly. Agri SA management is included in all negotiations and agreements with local agricultural unions. Joint communications platforms, such as newsletters, social media, and meetings, are also used to communicate to members on different levels. An example of a shared communication platform is the Agri Northern Cape Congress, where the project director of SKA SA updates the agricultural communities on relevant developments and meets with agriculture management. Agriculture-related radio and print media are also utilized as communication channels for individual farmers. In addition, Agri SA is involved in the reviews and development of supplementary telecommunication systems in the local, agricultural areas where mobile telecommunication will be affected.

The Agri SA MOU has implications at national, provincial, and local levels. Local farmers and agricultural businesses, referred to as farming unions, were organized by each town in the SKA-affected region. These unions created a forum called the Agriculture Action Group to engage on SKA-related matters as well as other developments in the area. NRF|SARAO meets with this forum once a month or as the need arises. Some of the topics discussed and negotiated include access control, supplementary communication systems, fence management, predator and invasive species control, land management in general, rural safety, and community development, to name a few. These meetings are attended by an Agri SA representative, or feedback is communicated to Agri SA. The local agricultural union is the communication link between NRF|SARAO and the farming community.

Negotiation, communication, and consultation with landowners is facilitated at different levels. When NRF|SARAO procured land, the core landowners were consulted individually and had the option to include Agri SA and other entities. This process has now been completed, and the land needed for the core of MeerKAT, SKA Phase 1, and SKA Phase 2 has been secured. The next phase of land acquisition for the three spiral arms shown in Figs. 8.2 and 8.5 earlier has commenced, and negotiations with these landowners are currently taking place. Some of the issues raised in these negotiations include land access and compensation for this access.

No more farms will be procured, but servitudes will be established where seven telescopes will be placed. This allows farming activities in the area to continue, with only 2 hectares fenced off where the telescopes and the solar plants that will power these telescopes are situated. The impact, therefore, will be minimal. However, in this process, collaboration and communication with local farmers are crucial, and the NRF|SARAO land manager works closely with the affected groups. In all cases, stakeholders are engaged and updated on the progress of the project as well as the technicalities relating to having the project infrastructure on their land.

## Giving a Voice to a Community: Local Public

The next level of engagement is with the public in the affected towns. Although the local municipalities act on behalf of the communities, some of these stakeholders do not support or consider themselves represented by the forum. Therefore, with the permission of the local government, NRF|SARAO also engages with the communities directly with a municipal representative present. While communities were initially only informed of SKA South Africa's plans without an opportunity to provide feedback, public opinion appears to be one of the main drivers of the social sustainability of the project, and therefore the community's contribution is now considered valuable to the project since it fosters feelings of ownership within the community. In addition, the project is hosted by these communities, and a number of NRF|SARAO staff hail from them. It is therefore very important that the local communities are well informed and that their voices are heard. The types of engagement that focus on the public include the following:

- **Community information and engagement sessions**: Each affected town is visited twice a year by NRF|SARAO management. The events include a keynote address by the project director to provide information on the project's progress and on social and economic developments. This is followed by a breakaway session where members of the community can pose questions or raise concerns with the experts including scientists and researchers from various NRF|SARAO departments.
- **Use of media**: Local community radio stations and newspapers as well as social media are used to create awareness and to communicate relevant events, SKA facts, and job opportunities in local communities.
- **Site visits**: Through the office of the NRF|SARAO stakeholder manager and communications division, interested community members are taken to the site to view the construction and progress of the project to provide context to the information. These site visits are coordinated and regulated, with specific days

allocated to site visits to protect the instruments. Community members are also included in the celebration of the achievement of all significant milestones, such as the completion of MeerKAT, which is currently conducting scientific observations.

- **Public consultation processes**: The project must comply with various legislative processes. For instance, it needs to consult the general public, and it needs to be included in municipal development plans that are approved through consultation with the general public. For instance, if the project needs a water license, the general public will be consulted.
- **Visits by senior officials from the Department of Science and Technology**: In the spirit of *Ubuntu* (a South African concept of humanity and the value of community), the Minister of Science and Technology, in collaboration with the NRF, visits the local communities from time to time. These events are referred to as ministerial *imbizos* and are true to their isiZulu origin as mass gatherings called by tribe leaders to give a platform and voice to the community. At these events, the Minister engages directly with the community to encourage unity, shared ownership, and support for the project.
- **SKA information center**: Since engagement events are spaced throughout the year, the NRF|SARAO has an information center in Carnarvon that the public can visit to access information about the project. This information center is operated by NRF|SARAO staff.
- **Access to the NRF|SARAO stakeholder manager**: The stakeholder manager interacts with the local communities on a regular basis to provide information and to engage with them on various public concerns as they arise. This interaction is especially important and necessary as it gives the project a human element or "face" to which the communities can relate and thus humanizes the project. The presence and visibility of NRF|SARAO officials also ensures that engagement activities are communicated effectively to community members.

The project sees the local communities as important contributors and partners. Therefore, input provided by the local communities from these various types of engagement is now included in all relevant strategic documents and plans relating to the project. Some examples are the selection of community development projects, communication strategies, the move from biannual information sessions to monthly community meetings in each town, and the communication strategy of the NRF|SARAO, to name a few.

## Reestablishing a Deteriorated Indigenous Knowledge System: San People

The land on which the SKA will be constructed in both Australia and South Africa has strong ties to indigenous heritage. In South Africa, the land was walked by the early ancestors of the San people, and there are still significant amounts of cultural heritage found on the land. This is similar to the Australian wing of the project, where the Wajarri people have a strong cultural heritage. Respecting these indigenous minorities and engaging with them is important to the SKA project. Therefore, agreements have been entered into with both the Wajarri people in Australia and the San people in South Africa. This section will focus on NRF|SARAO's engagements and activities with the San in the Karoo.

To truly understand why these groups of stakeholders are so important requires a backward glance. The San people of South Africa are among Africa's most intriguing people. Genetic evidence suggests that they are some of the most ancient people on Earth, having been around for the past 22,000 years (Barnard, 1992). For centuries, the San people of southern Africa have experienced colonial violence, ethnocide, and dispossession, which has pushed them into arid lands in the north of South Africa (Lee, Hitchcock & Biesele, 2002). By the beginning of the 20th century, the remaining San in southern Africa were to be found in the drought-prone areas of the Northern Cape, Southern Namibia, and Botswana. The San people's early ancestors, the !Xam, walked the land where the SKA will be hosted, but this group of people was driven from the land in the 1800s. Consequently, the San culture has almost disappeared from the area; some descendants remain in the area, but none are on the land now owned by the NRF. This is evident when one looks at the heritage found by Barnard (2002) on the land procured for construction of MeerKAT and the SKA. The San were driven from the land, first by settlers and later by the British government in the 1800s. The complexity of the San people's heritage and genealogy has been well documented by authors such as De Jongh (2012) and Bleek and Lloyd (1911).

In an attempt to right past wrongs, NRF|SARAO signed an MOU with the San people that is structured around the protection and promotion of the San culture and heritage, as well as the development of San youth. By taking the indigenous knowledge systems that focus on astronomy and cosmology seriously, NRF|SARAO ensures that this knowledge is not portrayed as unscientific or dissimilar to western science. NRF|SARAO engages in a variety of activities with the San to support these objectives, thus bridging the divide between ancient indigenous knowledge systems and one of the most advanced technology and astronomy projects in history.

### *Protection and promotion of San culture and heritage*

The protection and promotion of San culture and heritage are linked to the land. Although the land procured for the project in South Africa is legally owned by the NRF, the San people have a strong cultural and spiritual link to this land. Stories are told by the San about how the !Xam people once lived on this land. Furthermore, rock art in the area tells the stories of early San hunters, what they encountered on the land, how they understood the cosmos, and their first encounters with westerners.

Therefore, one of the first initiatives with the San after the agreement was signed was a ritual permission and land blessing ceremony that was held at the site in the Karoo. This ceremony was structured to ensure that the San people's story about cosmology was profiled and aligned with the story of modern radio astronomy. The land and the work conducted by NRF|SARAO were blessed by the San community, the land was cleansed of the past injustices, and ceremonial permission was given for the project to be constructed on the land. Throughout the event, there was a focus on the San's understanding of the creation of the cosmos. Through these rituals, the San emphasized their support for the project. Furthermore, ecology along with heritage studies was integral to the strategic environmental assessment required to construct the SKA. These studies included different groups that have heritage on site, which included the San. These studies entailed walking the land to identify cultural heritage, fauna, and flora. The findings were documented and shared with the San Council. From these studies, the need to conduct further studies on indigenous knowledge was identified, and NRF|SARAO, in collaboration with tertiary institutions and the San, will conduct these studies in the future.

The arts are an important aspect of cultural life for both the San people in South Africa and the Wajarri in Australia. The *Shared Sky* art project was developed to emphasize how these ancient tribes understand the creation of the cosmos and the similarities between these two ancient people. *Shared Sky* stems from a vision by the SKA to bring together Aboriginal Australian and South African artists under one sky in a collaborative exhibition to celebrate humanity's ancient cultural wisdom. This art is used internationally on different platforms, and the San people feel proud to have been part of this exhibition. It reflects the richness of the artists' ancestors' understanding of the world developed across countless generations through observing the movements of the night sky. This vision embodies the spirit of the international science and engineering collaboration that is the SKA project, bringing together many nations around sites in Australia and South Africa to study the same sky.

### Youth development
One of NRF|SARAO's commitments to the area where the telescope is being constructed is to youth development, including San descendants and other indigenous people from the area. The Memorandum of Understanding between the San and the NRF|SARAO emphasizes this commitment. NRF|SARAO offers bursaries to youth from the area. For school students from grade 9 to 12, bursaries are provided to attend a school in the area where mathematics and sciences are taught at a higher level. In addition, the project offers graduate and postgraduate bursaries in related fields and has developed a school for artisans in the Karoo with bursaries for qualifying students. Furthermore, the San people and the San Council are supported in terms of tourism and tourism-related projects where San youth are trained and serve as amateur astronomers and field guides. The San Council has also attended conferences and presented with SKA staff on issues related to San indigenous knowledge.

South Africa almost lost an interesting and unique part of its history: a part of history that is rich in archaic cosmology. An entire indigenous culture is being revived through NRF|SARAO's dynamic stakeholder engagement. Not only do the San people benefit, but South Africa is given another opportunity to secure a culture that is part of its rainbow nation. How appropriate for the world's largest and most advanced radio astronomy project to be built on the land where the very first stories of the cosmos were shared among the San around fires.

## Countering the Ripples of Negative Sentiment: Anti-Square Kilometer Array Groups

The project has received a lot of support from most stakeholders, and the positive impact on the country has been unprecedented due to the advancements and opportunities that the project brings to South Africa. However, the project has also met with some opposition. There was opposition to the procurement of land from people whose families had owned the land for generations. Furthermore, there was ethical and religious opposition to the project among people who regarded the SKA project as a threat to their understanding of the cosmos and how it was created. Consequently, three different but connected anti-SKA groups emerged from the local communities: Save the Karoo, which protested against SKA through social media and representation at SKA engagement events; Friends of the Karoo Against the Industrialisation of the Karoo, which employed social media campaigns; and Carnarvon Forum, which was closely associated with Save the Karoo.

As the most vocal group, Save the Karoo aimed its strategy at disrupting the processes of developing the SKA project and creating distrust among local communities. The group also launched a targeted media campaign using Afrikaans print media and local media to postulate that the project posed threats to local communities. The disruptive communication included many fictitious claims. For example, Save the Karoo claimed that SKA would procure nearly four times the amount of land it actually would, demolish two blocks of residential property, cut off telecommunications in the area, and collapse the local economy. In addition, the group propagated the untruth that SKA did not perform any environmental impact assessments and would commit ecocide, drain the area of almost all of its water reserves, and change weather patterns. Furthermore, the group maintained that SKA did not consult any local stakeholders and had no planned end date.

Save the Karoo was started in reaction to the land acquisition process and was aimed at halting the process. The conflict between NRF|SARAO and this group deepened with the declaration of the Astronomy Advantage Areas Act and the associated regulations that were proclaimed by the then Minister of Science and Technology. The SKA project's management team indicated that it would nonetheless engage with any local forums that wished to do so. Save the Karoo, in collaboration with a local church and some of the local farmers, then established a forum named the SKA Forum. This forum indicated in its constitution that it would be a voice both for and against the SKA. It was later found that this forum was purely an extension of Save the Karoo and it was therefore still an anti-SKA movement. NRF|SARAO management attended meetings with this group and provided written responses to all questions posed and statements made.

NRF|SARAO adopted a proactive approach to media and social media messages by posting facts about SKA and the development of the project as well as tailoring its communication in terms of research and information gathered from the local communities. The anti-SKA groups did not change their sentiment toward or understanding of SKA; however, as communities received information on the project and were provided with the opportunity to engage in dialogue, the disruptive communication from opposition groups was countered as far as possible, and the impact of these groups was minimized.

The insights gained from different research projects such as social media and media analysis, polls, and socioscientific surveys ensured that SKA could respond appropriately to the anti-SKA communications. This phase of the stakeholder engagement process ensured that the sentiment toward the project shifted from primarily negative to primarily positive. The process was assisted by NRF|SARAO's willingness to engage with these stakeholders directly as well as by giving other stakeholder groups the opportunity to raise any questions or concerns that resulted from negative communications. NRF|SARAO remains responsive to stakeholders' questions and concerns.

## Contributing to a Scientifically Literate and Engaged Society: Science Engagement

For NRF|SARAO, all stakeholder engagement incorporates science engagement and the communication of science and science content for the benefit of communities. Science engagement has drawn considerable attention in South Africa—so much so, that the National Research Foundation Act 23 of 1998 was amended in 2018 to prioritize it. So, what makes science engagement so important that it is incorporated into legislation? Government serves society and as part of the Constitution guarantees economic development. Human, social, and economic development have been inextricably linked to the advancement in science and technology. Science engagement with stakeholders creates awareness, and awareness often encourages community participation in the project as well as in science, engineering, and technology in general. This participation can lead to intragovernmental support, industry participation, critically engaged publics, and interest and subsequent inflow of new talent into the professional pipeline for science and technology. It is, therefore, a priority for the Government, and now a key focus area for the NRF, to engage stakeholders in science communication and science engagement activities. To promote a scientifically literate and engaged society, the NRF adopted an integrated science engagement model through which all NRF facilities, including NRF|SARAO, become involved in science engagement within their respective fields. Through collaborative efforts between facilities, government departments, industry, and global partners from over 40 countries, science engagement within the NRF is finding its foothold to improve the life of every South African for a better now and a more sustainable future.

Although science engagement is not the primary focus of some of the interventions described in this chapter, every SKA project interaction with stakeholders involved science communication to some extent. During every community meeting, the stakeholder groups were engaged not only on social and environmental factors but also on the importance of science and economic development. This was sometimes done overtly and at other times more subtly. The project also engaged those not directly affected by the project but who could potentially benefit from the engagement consistent with the South African Government's mandate regarding the advancement of science and technology. NRF|SARAO involved more than 58 scientists in a variety of science engagement activities, which included outreach workshops, career profiling, exhibitions, and public visits. The following provides examples of the types of engagement that NRF|SARAO would normally facilitate for some of the stakeholder groups not directly affected by the project.

The South African Association for Science and Technology Advancement (NRF|SAASTA) National Schools Debates Competition is one of the flagship projects that provides school students with an opportunity to develop their research, critical thinking, and information literacy skills as well as their ability to work as a team to present logical arguments. Among the objectives of the competition is to develop young science communication ambassadors through the students' work of researching and debating high-level scientific topics. In 2018, NRF|SARAO collaborated with NRF|SAASTA to host 13 workshops to prepare students for the debates with schools from the Northern Cape province. From minidebates held across South Africa, 10 teams were selected to represent their respective provinces at a national debate event. The topic for the provincial tournaments was hydrogen fuel cell technology and its importance as a possible solution for the energy needs of schools and society at large. The winning teams from each province gathered for a second round to debate whether the investment of public resources in megascience projects such as SKA can prepare South Africa and Africa for the fourth industrial revolution and contribute to sustainable development and the eradication of poverty. Through the collective engagements between NRF|SARAO and the schools, a Northern Cape school was awarded second place at the national finals.

In collaboration with science centers across South Africa, NRF|SARAO often hosts coding literacy workshop tours for high schools. Each workshop consists of an introductory session to explain what coding is and a hands-on session during which school students are taught basic coding using provided laptops preloaded with Scratch. In addition, school students are introduced to the SKA and MeerKAT, the careers in radio astronomy that need coding, and how the SKA project applies coding. In addition, robotics workshops were implemented in the Northern Cape, which led to SKA hosting the First Lego League Junior Kids training in Carnarvon. Training by NRF|SARAO content specialists introduces school students to robotics by allowing them to build various robotic models from Lego components. These school students then participate in the national First Lego League Junior Kids competition. Events such as these have had a significant impact on school students. Recently, a learner from a school where initiatives were hosted achieved a mark of 100% in physical science, while Carnarvon High School as a whole achieved a 75.4% overall pass rate. To this end, 15 school students are now enrolled at various universities, with 6 from the class of 2017. Without science engagement with NRF|SARAO, this may not have been possible.

Furthermore, NRF|SARAO and the University of Cape Town (UCT) have helped students by forming a partnership that enables teachers from the Northern Cape province to enroll for mathematics, physical science, and English courses at UCT. This opportunity is offered to teachers to enhance their subject content knowledge and teaching abilities and ultimately improve knowledge transfer to current and future school students. In addition, schools' governing bodies receive school management training. This provides only a small peek at NRF|SARAO's contribution to a better education system.

In addition, NRF|SARAO recruits bright, young, and enthusiastic university graduates for its Young Professionals Development Programme. These graduates are employed by the NRF|SARAO for 3 years to gain access to mentoring and acquire work experience, while they pursue a master's or doctorate qualification. These graduates work directly with scientists and engineers engaged in the project to develop their research capabilities, among other skills. NRF|SARAO also regularly collaborates with other African countries to host students from abroad. One example is the collaboration with the International Astronomical Union (IAU) Office for Young Astronomers (OYA) to host a science communication workshop for African postgraduate astronomy students in Addis Ababa, Ethiopia. This workshop is an extension of the IAU International School for Young Astronomers; it is a 2-week school that offers students selected by the IAU OYA with the opportunity to broaden their perspectives on astronomy through a series of lectures, practical exercises, observations, and exchanges facilitated by an international faculty.

In addition, NRF|SARAO participates in at least five major South African science festivals annually. Apart from erecting informative and interactive exhibitions that focus on astronomy and the project, representatives also engage with the public about bursary and career opportunities enabled by the project. However, activities do not stop at information dissemination. Several workshops are regularly hosted at public participation events. These workshops include a "Radio Frequency Interference Detective" workshop that introduces participants to the electromagnetic spectrum, radio waves, radio frequency interference, and the optimal conditions required for radio astronomy. The popular "Making a Pinwheel Galaxy" and "Make Your Own Spectroscopy" workshops are aimed at stimulating interest and excitement among the younger attendees.

## Conclusion: the Perspective of a Stakeholder Manager

Finally, South Africa, a developing country, is at the forefront of a complex science that could 1 day explain our origin and evolution and if we are alone. However, this pursuit does not come without a price: a multitude of stakeholders were, still are, and will be affected. It is easy to support a project when you consider the benefits from a global perspective, but will you be as eager to support it when you are directly affected? It is doubtful whether you will be as receptive when a multinational organization sends you a glossy pamphlet explaining the benefit of a project that you know will change your life in unknown ways.

This chapter outlines possible best practices in stakeholder engagement that took the NRF|SARAO years to develop. Is NRF|SARAO's stakeholder engagement framework effective? Yes, as most of its objectives have been met. Is this stakeholder engagement framework complete? No, not even close. The past has taught us that this effort will remain a continuous process based on trial and error. Why? Because this chapter does not express the full magnitude of the challenges experienced by the organization and its stakeholders before reaching a point of stability. It does not make provision for any instability; it does not address uncertainty; it does not consider the changing global political, economic, social, environmental, or technological landscape. It is based on past experiences. As each day passes, more challenges result in more experience and we adapt accordingly.

NRF|SARAO did not initially expect the profound effect it would have on its stakeholders, both positive and negative. However, NRF|SARAO does attempt to give a voice to its stakeholders; it no longer sends out information that it deems important and likely to result in positive sentiment. No. NRF|SARAO has stepped down from the podium to sit with the farmer whose brow is still dusty, the jobless individual nervously worrying, the businessman ready to invest, the child that is eager to learn. It now engages on the information that each of these stakeholders feels is important, in their context and from their perspective. The fundamental lesson learned from this process is that when an organization takes a moment to listen, to understand, to acknowledge, and to engage, it no longer stands alone; it fosters a community that unites, supports, and advocates. No organization can pretend to know what is best for its stakeholders if it has not given them a communication platform. It is only through engagement that a mutually beneficial relationship can be established, satisfying their needs and our needs. It is easier to have a community behind you, supporting you, than having one in front of you, opposing you. Is this not what stakeholder engagement is all about?

# References

Barnard, A. (1992). *Hunters and herders of southern Africa: A comparative ethnography of the khoisan peoples*. Cambridge: Cambridge University Press.

Bleek, W. H. I., & Lloyd, L. C. (1911). *Specimens of bushman folklore*. London: George Allen.

Cornelissen, J. (2014). *Corporate communication. A guide to theory and practice* (4th ed.). Los Angeles: SAGE.

Corntassel, J. (2008). Toward sustainable self-determination: Rethinking the contemporary Indigenous-rights discourse. *Alternatives, 33*(1), 105–132.

De Jongh, M. (2012). *Roots and routes: Karretjie people of the Great Karoo: Marginalisation of a South African first people*. Pretoria: UNISA Press.

Lee, R. B., Hitchcock, R. K., & Gisele, M. (Eds.). (2002). *Foragers to first peoples: The kalahari san today. Cultural survival quarterly* (Vol. 26 (1), pp. 18–20).

Overton-de Klerk, N., & Verwey, S. (2013). Towards an emerging paradigm of strategic communication: Core driving forces. *Communication, 39*(3), 362–382.

Roloff, J. (2008). Learning from multi-stakeholder networks: Issue-focussed stakeholder management. *Journal of Business Ethics, 82*(1), 233−250.

Sloan, P. (2009). Redefining stakeholder engagement. From control to collaboration. *Journal of Corporate Citizenship*, (36), 25−40.

## Further reading

Brochure, S. K. A. (2017). Retrieved from http://www.ska.ac.za/wp-content/uploads/2017/07/ska_adirc_fact_sheet_2017.pdf on 6 January 2019.

Collier, J., & Wanderley, L. (2005). *Thinking for the future. Global corporate social responsibility in the twenty-first century* (pp. 169−182). Futures 37.

De Beer, E., & Rensburg, R. (2011). Towards a theoretical framework for the governing of stakeholder relationships: A perspective from South Africa. *Journal of Public Affairs, 11*(4), 208−225.

Holmström, S. (1996). *An intersubjective and social systemic public relations paradigm.* Doctoral dissertation. Roskilde, Denmark.

Independent Evaluation Office. *The big data revolution for sustainable development.* Retrieved from http://gefieo.org/sites/default/files/ieo/ieo-documents/SDG-Bigdata.pdf on 6 January 2019.

Porter, M. E., & Kramer, M. R. (2011). *Harvard business review: Creating shared value.* Retrieved from http://hbr.org/2011/01/the-big-idea-creating-shared-value/ar/pr.

Rasche, A., & Esser, D. E. (2006). From stakeholder management to stakeholder accountability. *Journal of Business Ethics, 65*(3), 251−267.

Walters, F., & Takamura, J. (2015). The decolonized quadruple bottom line: A framework for developing indigenous innovation. *Wicazo Sa Review, 30*(2), 77−99.

# Crowdfunding for Space Science and Public Engagement: The Planetary Society Shares Lessons Learned

# 12

**Jennifer Vaughn, Louis D. Friedman**

It was 2:30 a.m. at Florida's Kennedy Space Center when the third SpaceX Falcon Heavy rocket roared into space on June 25, 2019. The spectacular moment excited space fans from all walks of life, most staring upward and cheering, some with cameras, some with mouths agape, and many with tears in their eyes (see Fig. 12.1). It was a moment that looked like many others, but there was something different about this launch for many of these launch viewers: these space fans were watching "their" spacecraft leave Earth. Onboard that particular launch vehicle was a small solar sail

**FIG. 12.1**

Crowd watching LightSail 2 launch. *N. Baraty and The Planetary Society. (2019).*
*Retrieved from https://www.planetary.org/assets/image/society/baraty-lightsail-launch.html.*

Space Science and Public Engagement. https://doi.org/10.1016/B978-0-12-817390-9.00008-7

spacecraft called LightSail 2 that was funded entirely through crowdfunding. Over the 10-year history of the project, 49,426 people from 109 countries gave more than $7 million to make LightSail a reality. Some gave $1, most gave tens to hundreds of dollars, and a few gave substantially more. It was the mobilized crowd—the collective effort from diverse people with a shared passion—who were the proud "parents" of LightSail. They beamed with pride and excitement as they watched LightSail begin its adventure in space.

LightSail is a shining example of the promise of crowdfunding for space projects. If, in 2019, there was an example of nearly 50,000 people crowdfunding a $7-million space project, what might be possible in future decades? Imagine what 100,000 people or 1 million people could do. Could we one day see crowdfunded missions exploring other worlds? For The Planetary Society, the organization behind LightSail, crowdfunding space projects has been a long, challenging, and deeply fulfilling journey. By sharing some of our experiences and lessons learned along the way as a long-time leader and a founder of The Planetary Society, respectively, we hope the space industry can work together to unlock the potential of crowdfunded space projects—and thereby directly engage more people than ever in space science.

## The Planetary Society and Crowdfunding

Crowdfunding, as a term, is relatively new. The word generally refers to the process by which groups of individuals come together to support a project, product, activity, or event by contributing money via the Internet. The larger idea behind the word has a much longer history and does not require the Internet, although the Internet has substantially contributed to crowdfunding's popularity and capacity growth.

To best tell The Planetary Society's story, which began long before the Internet, we will use the term crowdfunding to refer to individuals pooling private financial resources—in this case, to advance space projects. Crowdfunding is built into The Planetary Society's DNA. At the end of 1979, our cofounders Carl Sagan, Bruce Murray, and Louis D. Friedman (one of the authors of this chapter) started The Planetary Society as a grassroots membership organization to prove and harness public support for planetary exploration (see Fig. 12.2). At that time, the United States had dramatically scaled back planetary exploration. Carl and Bruce were actively advocating for more exploration. As they did so, they recognized decision-makers were using perceived public apathy to justify defunding US planetary exploration.

Having worked with the Mariner, Pioneer, Viking, and Voyager missions, Carl and Bruce had ample anecdotal evidence supporting public enthusiasm for planetary exploration, but it was difficult to prove the breadth and depth of this public support. This challenge led Carl and Bruce to reach out to Lou, an engineer at the Jet Propulsion Laboratory who was finishing a year in Washington D.C. as a Congressional Science Fellow on the Senate space subcommittee staff. The three began to explore the idea of a space advocacy organization modeled after popular environmental and

FIG. 12.2

The founders of The Planetary Society. *The Planetary Society. (1989).*

*Retrieved from https://www.planetary.org/explore/projects/vom/the-founders-of-tps.html.*

science-related organizations, such as the Sierra Club and the Audubon and Cousteau Societies.

The founders identified key tenets to form the foundation of the organization. A few of these institutional pillars set up The Planetary Society to be especially well suited for crowdfunding. Most importantly, the organization needed to be broad and open to the public. In an August 1979 draft description for the still-unnamed organization, Lou Friedman wrote, "As space exploration has no single rationale, neither does it serve a single constituency. Those who care about its conduct come from many walks of life and hold many outlooks on why and how space exploration should be conducted" (Friedman, 1979). Lou went on to write that the goal of this nascent organization would be to "bring together the various constituencies and to provide a public opportunity for participation in and support of the continuing exploration of space" (Friedman, 1979). This commitment to developing a broad and active community gave The Planetary Society a built-in base for project-specific fundraising. The founders also set the goal of creating public—private partnerships. Their vision was to create an organization that would be symbiotic with governments and space industry players, neither dependent on nor wholly independent from government and commercial activities. In time, this positioning allowed the organization's members to participate in myriad space projects with government space agencies, academics, and private space companies. All three founders were passionate about providing high-quality communication with the public. They believed that telling the stories and sharing the results of planetary exploration would be the best way to maintain and grow public interest over time. Their commitment to developing in-house communications expertise greatly enhanced the community's connection with the projects they supported. Members could track the impact they were having on space exploration.

In just 6 months, our organization—now called The Planetary Society—was ready to incorporate. The organization's original articles of incorporation state that The Planetary Society's purpose is to:

**(a)** enhance public awareness of the results from the exploration of the planets and the search for extraterrestrial life;
**(b)** to distribute and communicate among interested groups and individuals the latest findings and discoveries about the exploration of this and other solar systems and the search for extraterrestrial life;
**(c)** stimulate and facilitate new experiments, analyses and syntheses on the exploration of planetary systems and the search for extraterrestrial life (The Planetary Society, 1979).

The final bullet reflects the founders' foresight to use voluntary contributions to fund space projects. In the first issue of The Planetary Society's member magazine, *The Planetary Report*, Carl Sagan wrote: "If we are as successful as at least some experts think we are likely to be, we may be able to accomplish not only our initial goal of demonstrating a base of popular support for planetary exploration, but also to provide some carefully targeted funds for the stimulation of critical activities" (Sagan, 1980, p. 3).

Just a few months after those words were published, we tested the theory. In June 1981, we mailed our first project-specific fundraising letter to all Planetary Society members. (In pre-Internet, pre-telemarketing days, money was raised by direct mail, and crowdfunding attempts were called "junk mail.") That first effort generated more than $70,000 (the equivalent of more than $200,000 today after adjusting for inflation). Our projects started off small and very specific: from that first fundraising effort, we funded US scientists to travel to Tallinn, Estonia, for an international search for extraterrestrial intelligence (SETI) conference (after NASA funding for SETI was cut off), gave seed funding for a high-resolution hardware spectrometer to help process SETI data, and funded pioneering researchers hunting for exoplanets.

As our community responded so well to crowdfunding opportunities, we began seeking regular and more ambitious projects. One of our hallmarks when it comes to crowdfunded activities is our focus on partnerships and relationships: some with government, some with commercial companies, and many with universities. Our members supported researchers, enabled science meetings, contributed to hardware and technology development, and eventually developed payloads that would fly to other worlds. Each time we stretched our crowdfunding capacity, we seemed to raise the bar and set new expectations from our members. Lou Friedman remembers a meeting with Bruce Murray and Carl Sagan in which Carl wondered if one day we might be able to fund our own space mission. With LightSail, the current apex of our crowdfunding capacity, the Society did just that.

In addition to project-related work, our members supported program activity, such as educational initiatives and the Society's advocacy and space policy efforts. Diverse crowdfunding ensured The Planetary Society could keep a long-term view and be an independent voice advocating for future exploration. Even while pioneering the use of private funds in space exploration, the Society has remained a strong advocate and supporter of government space ventures. The opportunity for meaningful public involvement in space is what led The Planetary Society to a crowdfunding approach, and the success of crowdfunding led to greater independence for the organization.

## Developing Crowdfunding Capacity

For The Planetary Society, developing crowdfunding capacity has meant prioritizing two things: developing our community and providing opportunities for the community to take action for space. Our story began with an unlikely source: John Gardner, the founder of Common Cause and former Secretary of the United States Department of Health, Education, and Welfare. Gardner, whose main interest was in creating and developing constituencies for the benefit of society, recommended The Planetary Society to a like-minded philanthropist/hotel owner as a worthy new organization building a citizen group for planetary exploration. The philanthropist provided seed funding to The Planetary Society on the condition that we engaged a direct mail contractor to conduct a membership drive and raise money for worthy projects that would fund scientific exploration. The Society leveraged that funding to become the fastest growing membership organization in the world in the decade of the 1980s.

The essence of crowdfunding had been explained to The Planetary Society well before the term had been invented. Lou Friedman remembers Gardner explaining that members want to be part of planetary exploration: they want projects that they create, own, and make happen to advance the exploration of other worlds. Essentially, the Society created an opportunity for individuals to participate in space. Before the popularization of the Internet and the advent of social media, community development took place in person and through the mail. To nurture this growing community, we produced a member magazine with in-depth articles about space, *The Planetary Report*, and opportunities to take action; we held public events; and we sent regular mailings to inform our members about our projects and activities and opportunities for member action through surveys, petitions, and crowdfunding. Keeping members engaged through information was a key to providing them with a sense of "ownership" in the projects.

The advent of online communications—email, blogs, web comments, user groups, and social media—accelerated our ability to engage our community and simplified two-way communication so we heard more, and heard more regularly, from our supporters. Now, in addition to tending to member relationships, we are actively cultivating a much broader audience of space fans. Developing an active community is an ongoing commitment. These are relationships, and they need attention just like all relationships do. As mentioned before, community organizing and engagement is baked into our *raison d'être*, so these actions support both our mission and our business model. For others who are developing communities and using crowdfunding methods, a more pared-down approach to community development might make more sense, but community engagement is still essential to successful crowdfunding at any level.

This collective group of space enthusiasts is motivated to donate time, effort, and money to ensure space exploration continues. In our experience, community interest outweighs current opportunities to participate in space. Thousands, tens of thousands, perhaps millions of people want to do something with their passion for space. Without a steady stream of meaningful activities for these people, the potential of crowdfunding for space will remain untapped.

## Why Individuals Give

Nonprofits regularly scrutinize what inspires people to open their wallets to support particular causes. Donors—traditional philanthropists as well as occasional crowdfunders—are similar to investors: they are voluntarily giving their money to fund someone else's idea with the expectation they will receive a return on that investment. For donors to space projects, what is that return? Over the decades of working with Planetary Society members and donors, we have identified some key categories of "return" space enthusiasts seek.

### Impact

Financial supporters of space projects are true believers in the dream and promise of space exploration. This community does not just like space: they are passionate, and they want to do something with their passion. They want to know that, as individuals, they are positively impacting future exploration and discovery. Their return on investment is results.

### Pride

These passionate community members are often quite proud of their association with space projects. For many, contributing funds (at any level) gives them a feeling of "ownership" in activities that they are passionate about but would not otherwise be accessible to them. Membership in a group or participation in a particular campaign can reinforce their role as insiders, as explorers. Space exploration is an extraordinary and distinctive achievement of humankind. The achievements are scientific (new knowledge), technical (reaching new distances and new worlds), and cultural (changing our understanding of ourselves and our world). To be part of that—indeed, to help enable such advances—provides a profound reward to many donors.

## Recognition

The opportunity for recognition can help move a person to donate. To date, the most money raised for a space project via the online fundraising platform Kickstarter was for Planetary Resources' Arkyd Space Telescope. Arkyd offered funders of $25 or more the chance to see their selfie (or other image of choice) in space. In the end, the project was scrubbed (more about that later), but the idea of seeing themselves in space was a powerful return for supporters. Other forms of recognition—names listed in public forums or digitized and sent on spacecraft, digital or physical "badges," and special products only accessible for insiders—can all be motivators for giving.

## Access

In some cases, supporters of space projects are looking for special experiences. Perhaps their participation in a crowdfunding campaign gives them a chance for a behind-the-scenes tour or an opportunity to meet an influential person working in the space industry. Some crowdfunding platforms, such as the company Omaze, are built on experiences such as these.

## Stuff

Over the years, we at The Planetary Society have learned that stuff—t-shirts, pins, signed prints, etc.—rarely motivates someone to decide to donate, but it can motivate someone to give at a particular level once there is desire to support a project. Special gifts, especially those that are exclusive or branded to the project, can also help donors feel connected and recognized. Crowdfunding platforms such as Indiegogo and Kickstarter use this type of "rewards" structure to help interested donors choose funding levels that fit their interests and the project goals.

---

# Successful Campaigns

Why do some crowdfunded projects receive full funding in days, while others never come close to their funding goals? What compels a donor to give a first gift, and—even more importantly—what inspires that person to give again? While there is no guaranteed path for a fundraiser to follow, proper planning and preparation are good places to start. We have learned there are some common elements to successful crowdfunding campaigns, whether they are one of our special funding appeals to our members or broad online approaches to crowdfunding.

## Establish Credibility

Why should potential donors trust us to use the money effectively and efficiently and deliver the "return" they seek? For some, building credibility might mean showcasing historical successes, proving that they have delivered in the past so they are a good bet for the future. Others may employ or point to influential people who endorse the project. Even building up a record of failed and successful crowdfunding campaigns using online platforms can boost credibility. Perhaps most importantly, other funders boost credibility. If other donors support the project, it is easier for new donors to take a leap of faith, which leads to the next important element: momentum.

## Create Momentum

Momentum also generates credibility. Donors are more apt to invest in a project with a high probability of being successfully funded. The majority of crowdfunding these days is done online, and it is common for all donors and potential donors to be able to see how well a campaign in doing toward its funding goal. A slow start can doom a campaign. Before even announcing the goal, we like to have some funders lined up and ready to donate in the first few hours (or even the first few minutes) of a campaign. Similarly, it is helpful to have promotional partners already lined up to help spread the word when the campaign goes live. Crowdfunding consultants often suggest setting a more modest financial goal that should be possible to hit in just a few days. Once the initial goal is set, it is good have some "stretch" goals ready to move toward the full funding goal. A note of caution: make sure the initial funding goal is enough to fund the entirety of the project, including the costs associated with the crowdfunding campaign (e.g., transactions fees and fulfillment of any "rewards" that are offered). No one wants to end up with a fulfillment obligation and not enough funding. Setting the right initial funding goal is an essential step in developing a successful campaign and deserves ample time and thought.

## Be Specific

As noted earlier, donors want their contributions to make a difference. Taking time to think about the specifics that need funding can help donors visualize their potential impact. Let us say a hardware project needs $100,000. Is it possible to break down the cost to more bite-sized pieces? Do we know the funding needed for each component or each system or each person doing the work? That kind of specificity can help donors visualize their impact. Even with research projects, it is possible to add specificity. In 2013, The Planetary Society raised funds for the search for exoplanets in the Alpha Centauri system. At the time, the cost for one night of observing at the 1.5-meter telescope at Cerro Tololo Observatory in Chile was $1650. We shared the specific cost with our members, and we broke down an hourly rate as well. That campaign raised $96,943, of which $32,063 came from donors giving amounts specific to the hourly and nightly rates.

## Know Your Story, Show Your Story

Of course, projects vary, and getting down to granular specifics is not always possible. What is always possible is good project planning and good storytelling.

Like all investors, donors want to see budgets and timelines and updates on progress. Contemporary crowdfunding platforms favor strong stories, and the audiences have high expectations for engaging videos, friendly prose, infographics, and evocative images. Using such a mix of modalities can also help walk the thin line between telling a thorough and transparent story while not overwhelming the potential donor.

## Communicate, Communicate, Communicate

As crowdfunding has become more popular, the public expectation for regular communications has escalated. Supporters expect regular project updates and quick answers to their questions.

To cultivate a crowdfunding community, we at The Planetary Society place a high priority on communication and gratitude throughout the campaign and throughout the project. We are fortunate because our business model fully supports a communications team. For example, for our LightSail project, we had a journalist assigned specifically to cover project development. Since 2014, he has produced more than 120 behind-the-scenes stories about the project. Nevertheless, we have been criticized for not communicating more (not because our donors have read all 120 stories but because communication is asynchronous and when they are looking, we might not be posting). Public expectations are high.

In most contemporary crowdfunding campaigns, there is also merchandise to fulfill. (Remember the "rewards" mentioned earlier?) Supporters expect a level of customer service when it comes to these rewards. In today's world with overnight or even same-day shipping, managing expectations can be challenging. Chances are if you are raising funds for a space project, you may not also be an expert online retailer. That is okay, as long as supporters know what to expect.

Supporters expect transparency. For us, this means if something is not progressing the way we hoped, our community of investors deserves to know what we know. Not surprisingly, responses to setbacks can widely differ, from messages of support to cruel outbursts. We have found that transparency is always the best practice for long-term relationships, even if short-term interactions are dicey for a bit. For all of these community relations needs—project updates, customer service, and potential crisis management—fundraisers should expect to put resources toward communications.

## Follow Through

Over the decades, hundreds of thousands of people have put their trust in our projects to help fulfill something within them. It is our job to do everything we can to deliver on the myriad expectations. Deliver the physical and experiential rewards, recognize and credit the donors, provide points of pride for the donors to remind them of their essential role, and—above all—complete the project. All of these aspects are in our control. If we have planned and allocated appropriate resources, we can deliver. There is an interesting example of a successful crowdfunded space project whose parent company chose not to complete the project. Cited earlier for the most money raised in Kickstarter for a space project, the Arkyd space telescope raised $1.5 million from more than 17,000 supporters. About 3 years after their Kickstarter success, the parent company, Planetary Resources, announced to their backers that they would not be following through with the project and offered supporters a full refund (Lewicki, 2016). While there were plenty of backers expressing disappointment and frustrations on the Kickstarter page, many of their backers did not want a refund. Instead, they wanted Planetary Resources to use their funds for other projects.

The ultimate return on investment is the part we have less control over: results. With space projects that are—for the most part—experimental, we cannot guarantee particular results. Planetary Society members have been financially supporting SETI since 1981, but no one has yet detected signals from an extraterrestrial civilization. Similarly, nearly two decades ago, Planetary Society members funded a hunt for Vulcanoids, a hypothetical population of asteroids that orbit the Sun. Our members enthusiastically supported a suborbital flight that resulted in no observable Vulcanoids. We delivered on the project but did not—could not—deliver on the results. When we have not been able to deliver results, we have doubled down on pride as a return on investment. Our supporters are actively exploring and watching the scientific method play out. They are fueling experiments, and the results—positive, negative, or null—benefit future efforts. Our members began funding the search for exoplanets in 1981, and we had no planet confirmations for more than a decade. Now we have thousands of confirmed exoplanets, and our members can be proud of their role in advancing early searches.

## Listen to Feedback

We have seen the connection between listening to public feedback and growing crowdfunding capacity. One of our phrases around the office is "members vote with their wallets." It is a reminder that our members fuel all that we do. Planetary Society staff or leadership might think we have an exciting project, but it is the members who will verify if it is exciting or not. We had an experiment to test the viability of various forms of microbial life on an interplanetary mission. We raised the majority of the funds we needed from one enthusiastic member, but when we took the project to the broader community for crowdfunding, we did not come close to our funding goal. We tried one more campaign, and we received the same result, so we stopped all crowdfunding efforts for the project. Knowing that the project did not have the same level of popular support that other projects did helped us because we were able to change our approach. We worked with the one funder who liked the project and secured the funding we needed, and we took other, more popular, ideas to the crowd.

Of course, in today's communication-heavy world, voting with one's wallet is just one of many ways to gauge the popularity of a project. Watching communications channels—our own social media and email lists, other online conversation threads, media pickup—helps us hone our activities to better match community interest.

## It Costs Money to Raise Money

All forms of fundraising cost money. Some methods are more cost-efficient than others. Raising relatively small amounts of money from large groups (exactly what we are talking about in this chapter) is not one of the most cost-efficient fundraising methods. For The Planetary Society, however, crowdfunding is tied to our mission of empowering individuals to advance space exploration. When people get involved by giving to a project, we are fulfilling our mission through that participation, so crowdfunding is an extremely good fit for our organization.

In the 1980s, crowdfunding was done in the mail, so we had expenses for printing and mailing as well as for the direct mail experts to help write and design the mailings. Now that most crowdfunding has moved online; the types of expenses have changed but the cost of fundraising has not necessarily gone down. For online campaigns using platforms such as Kickstarter and Indiegogo, a percentage of the funds raised goes to the platform. When raising funds online, all the payments are via credit card, so there is a 2%–3% fee to factor in. Contemporary campaigns often offer rewards, some of which may require design, manufacturing, and mailing. Lastly, outside partners are often necessary to help with product fulfillment, communications, or other administrative needs. While it costs money to raise money, some of the costs are not fundraising expenses; rather, they are costs to disseminate information and engage the community being served. Crowdfunding campaigns are educational, providing background on project plans and updates on project progress.

The Planetary Society is very interested in seeing crowdfunding for space projects develop and grow. We think that one way to do this is to match strong project ideas with strong community builders and crowdfunders. This way, project experts do not need to be community builders, and community builders do not need to be project experts. For The Planetary Society, this means beginning to conduct open calls for proposals from the space community. If other organizations and companies joined in this or similar efforts to "match-make" communities with space projects, we think that crowdfunding capacity could grow.

## Flight Opportunities

Throughout our 40 years, we have encountered hundreds of thousands of space fans interested in supporting experiments, research, hardware development, testing, and even programs like advocacy and education. From our experience, nothing ignites public enthusiasm, pride, and support like a space flight opportunity. The Planetary Society is quite proud to have been a trailblazer for crowdfunded flight hardware projects. Our first opportunity came in the early 1990s, when the US government was not investing much in the planetary program. At that time, the Soviet Union had a series of planned Mars and Venus missions as well as a mission to Comet Halley. The Society had been building relations with Soviet scientists and engineers working on planetary missions, and those relationships led to our members funding part of a Mars balloon proposal, in cooperation with Soviet and French space agencies. This marked the first time a citizen-based organization had an official role on a planetary mission. Although the Mars balloon never flew, the work on it led directly to the Society being involved in Mars rover testing. Our international work and advocacy for Mars rovers influenced NASA to plan rovers in their program later in the 1990s. The impact of the Society's work raised the bar for our members and set a precedent for future relationships that we built with other space agencies, including NASA.

As noted earlier, public engagement is very important in citizen-based science and crowdfunded projects. As part of our involvement with Mars rovers, the Society promoted opportunities for our members to "fly" their names to Mars. The first flight project where this could be done was NASA's Mars Pathfinder, which launched in 1996. We collected many tens of thousands of names, which were then copied by microlithography on a microdot to be placed on the spacecraft. Scientists noted that simple materials on the microdot could record the radiation received on the lander while on Mars. This passive, low-cost public engagement project added scientific value, although there would be no way to provide an interface to send the data back to Earth. The radiation record would stay on Mars until some future spacecraft (or astronaut) could retrieve it. This experiment with the radiation sensors and the names of Society members was known as the Microelectronics And Photonics Experiment (MAPEx), and it is still on Mars today on the Pathfinder lander. Financed by the Society and flown to Mars by NASA, it was the first privately funded payload to another world and the first crowdfunded contribution to a NASA project. The Planetary Society went on to create other privately funded payloads to go to other worlds, including a digital "library" sent to Mars for hoped-for eventual discovery by future human explorers. This library, called Visions of Mars, was first created for Russia's ill-fated Mars '96 mission but then got a second chance with Mars Phoenix, which landed safely in the north polar region of Mars in 2008.

With flight opportunities comes risk. Anyone working in the space industry knows too well that space is hard and things go wrong. Helping supporters cope with these losses is an important part of developing a community that will support space projects. The Planetary Society has contended with a number of losses over the years. We shared one of these losses with thousands of our members. Planetary Society members had, over years, funded the development of the first microphone to fly to Mars (see Fig. 12.3). The Planetary Society–led team used relatively inexpensive "off-the-shelf" components and successfully integrated the microphone into an instrument aboard Mars Polar Lander. In December 1999, thousands of our members and supporters gathered in Pasadena, California, to watch the landing and witness their instrument send back the first sounds from Mars, but Mars Polar Lander crashed on the Martian surface.

Such moments are excruciating for all involved, especially the spacecraft team who put years, maybe decades, of effort into designing, building, testing, and developing operating plans for a spacecraft that would never get the chance to explore.

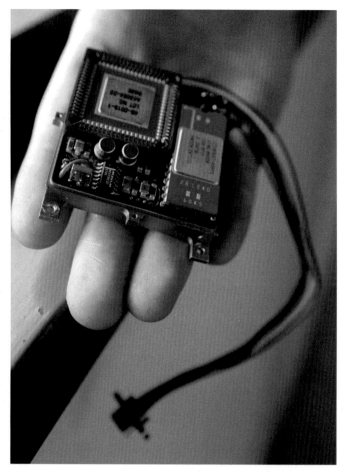

**FIG. 12.3**

The flight spare of the Mars Microphone that flew on Mars Polar Lander. *R. Weiner. (1998).*
*Retrieved from https://www.planetary.org/explore/projects/microphones/images/the-mars-microphone.html.*

The feeling of mourning is profound, and we recognized that our supporters—as extended "family" of the spacecraft team—were experiencing at least a taste of that grief. It was our job to help the community process and heal from the disappointment, and part of that job was to keep pushing forward with member projects. Like those in the space industry, our community bounced back and was ready to help with the next opportunity. Like space professionals, space enthusiasts are willing to accept risk and suffer disappointment to explore and discover more about our solar system and beyond.

For The Planetary Society, our most profound setback was the loss of our first solar sail spacecraft, Cosmos 1. At the time, the spacecraft—a collaboration between The Planetary Society, Russian space agencies, and the entertainment company

Cosmos Studios—was the largest and most complex project in the organization's history. Cosmos 1 would have been the first solar sail to be tested in space (see Fig. 12.4). The spacecraft launched in June 2005, but the rocket failed, and Cosmos 1 never made it to orbit. The loss was devastating to the leaders, staff, members, and supporters of The Planetary Society.

Once again, a project—although a failed one—raised the bar for our members and our organization. We had just rallied thousands of people to support a mission; what should we do next? Just 2 months after the launch failure, we took that question to our members and asked them what they wanted us to do: should we try another solar sail mission? The community response was strongly in favor of trying again. We received hundreds of supportive messages, such as this one from Nicholas P.: "Keep up the faith, you guys …. Even if the worst occurs, be proud of the fact that you were audacious enough to try. I have never been prouder to be a member of the Society than today." Through a combination of crowdfunding and backing from our major project sponsor, Cosmos Studios (whose chief executive officer is Ann Druyan, Carl Sagan's widow and collaborator), we also exceeded our first fundraising goal of $250,000 to help the organization explore different designs and launch opportunities. The strong member support led us to try a new approach,

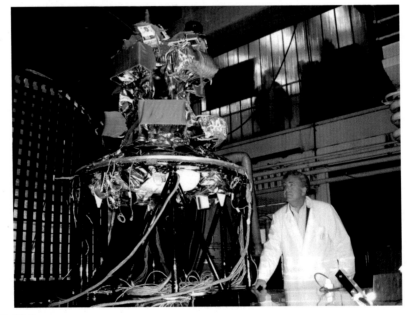

**FIG. 12.4**

Louis Friedman with the Cosmos 1 spacecraft. *Lavochkin Association and The Planetary Society. (2004).*

*Retrieved from https://www.planetary.org/multimedia/space-*
*images/spacecraft/louis-friedman-with-cosmos-1.html.*

pairing the innovative new developments of CubeSats with solar sail technology. This time, the organization would lead (and fund) the entire project. The project would become LightSail: citizen-funded flight by light.

## The LightSail Story

To date, LightSail is the most complex and most expensive project The Planetary Society has undertaken. A project of this scale led us into uncharted territory: at the time the organization committed to pursuing LightSail, we could not fully understand the total time and financial resources needed for success. A 10-year process of development and fundraising began where Cosmos 1 left off. Our members, still disappointed about the loss of Cosmos 1, rallied behind a new idea for a solar sail, one that would use a lower-cost CubeSat for the spacecraft bus. CubeSats are small, standardized spacecraft popular with universities as well as space agencies for flights in Earth orbit, but they have lacked a flexible interplanetary propulsion technique like solar sailing. Our 3-unit CubeSat, which measured $10 \, cm \times 10 \, cm \times 30 \, cm$ (3.9 inches $\times$ 3.9 inches $\times$ 11.8 inches), contained a stowed solar sail that, when deployed, measured 32 square meters (340 square feet) (see Fig. 12.5). The CubeSat design would be measured in millions of dollars rather than tens of millions of dollars or more for spacecraft similar to Cosmos 1. With the lower price tag, we were inspired to fund the entire mission through individual contributions.

FIG. 12.5

Artist's concept of LightSail 2 above Earth. *J. Spradling and The Planetary Society. (2015).*
*Retrieved from https://www.planetary.org/multimedia/space-images/spacecraft/ls2-earth.html.*

Between the end of Cosmos 1 and the announcement of LightSail, our members gave about $500,000, and we received $250,000 in seed funding from a cable TV network in cooperation with Cosmos Studios. These early funds went to exploratory activities such as evaluating designs for solar sail spacecraft. Late in 2007, in a handwritten note scrawled on the back of a fundraising letter, one of our members expressed interest in doing more to accelerate our solar sail project. The member, anonymous by request, pledged $1 million for a solar sail mission. We received this generous gift in 2009 and the LightSail project officially began. The $1 million allowed us to get to work, but we still needed to raise a lot of money to reach our goal. Over the following years, our members contributed hundreds of thousands of dollars and the anonymous donor pitched in more as well. This cumulative support was enough to ready our craft for a sail deployment test in Earth orbit.

The LightSail experience—like earlier experiences with Mars rover testing, the Mars microphone, and Cosmos 1—emphasizes another important aspect of crowdfunding: a good crowdfunded project leverages opportunity. The opportunity can be external—such as an anonymous donor or an Internet bubble creating a donor base—or internal—such as the application of small satellite technology. A good project idea paired with opportunity improves chances for crowdfunding success.

In May of 2015, our test spacecraft—LightSail 1—launched on an Atlas V rocket. The flight allowed our team to test the communications systems, practice operations, and try out solar sail deployment. We anticipated this milestone would be a big moment for our members, our broader audiences, and the media, and it was. The launch resulted in 10 million social media impressions on our own channels and 2.5 billion media impressions and garnered coverage in *The New York Times*, *The Washington Post*, Reuters, The Verge, Engadget, Aljazeera, GeekWire, *Popular Science*, CBS News, NBC News, CNN, and International Business Times. In 2015, we were still seeking funding for our full orbital flight (LightSail 2), so we did our first-ever Kickstarter campaign timed with the test flight of LightSail 1 to leverage our increased visibility. We set our initial goal at $200,000, which we hit in less than 2 days. We then sailed past stretch goals of $325,000, $450,000, $550,000, and $1 million. When the campaign closed, we had raised $1,241,615 from 23,331 backers—the highest number of backers to support a space project on Kickstarter.

The successful Kickstarter campaign meant we had the resources we needed to complete LightSail 2. What we did not anticipate was how long we would have to wait for a launch. Our launch vehicle was the brand-new SpaceX Falcon Heavy. We had thought we would launch as early as 2016, but in the end, we launched in June 2019. The 3-year delay meant increased operations costs as we kept our team prepared for an imminent LightSail 2 flight. While we waited, the team had the opportunity to continue testing the spacecraft and making improvements to the flight software, all of which added costs. Once again, we found ourselves needing to raise funds. Instead of returning to our members or the Kickstarter audience, we chose a different technique. We worked with the online fundraising platform Omaze. An Omaze campaign is similar to an online version of a raffle.

The donor buys a chance (or many chances) to win a prize. With Omaze, these prizes are experiences, usually with a celebrity. For LightSail, we knew we could offer a chance to view LightSail launch on a Falcon Heavy with Bill Nye, the Science Guy, who also happens to be The Planetary Society's chief executive officer. We ended up running two campaigns, and together, they raised about $250,000.

Lastly, we had two additional generous donors who supported LightSail with hundreds of thousands of dollars. All of these efforts led to the happy moment of launch as LightSail supporters from around the world cheered as their spacecraft started off on its journey. One week after launch, spacecraft operations began, the sails deployed, and our team successfully used the Sun's photons to change the orbit of our spacecraft. We declared mission success on July 31, 2019. As of this writing, the spacecraft is still operational in Earth orbit.

Like with the test flight, we anticipated a big media moment with LightSail 2. We were pleased to see that, along with stories about the preparation, launch, flight, and mission success of LightSail 2, the story of crowdfunding also got the media's attention. Inverse.com wrote: "… it's also notable just what this kind of innovation means for the future of crowdfunded space exploration. Though some previous attempts at this have been unsuccessful (or even downright painful), this high-profile work by Nye and the Planetary Society is the best proof yet that space isn't just the domain of governments or billion-dollar corporations. Space really is for everybody" (Manning, 2017).

And *Popular Mechanics* reported: "At a time where space travel has become increasingly big business and militarized, LightSail 2 is something of an anomaly because the seven million dollars needed to get the project in the skies was entirely crowdfunded over a decade" (Grossman, 2019).

Visionary supporters of LightSail 2 had much to be proud of. Here is a sampling of what we heard from supporters:

*SO EXCITING!!!!!!!!!!!!!!!! It was totally worth staying up all night to see another successful SpaceX launch and all those amazing satellite payloads being deployed, but man, when LightSail was released … it felt EPIC! Fingers crossed that 40,000+ people are walking around with sore faces from smiling so much over the next couple of months … GO LIGHTSAIL2!!!*

**Dave W.**

*A proud moment to know members like us have helped towards such a grand moment in space history!*

**Danny C.**

*So proud to back this. So proud of every person who made this possible. Surf that Sun.*

**Dana S.**

*So proud that my support could be a small part of this huge achievement. Bravissimi to everyone involved.*

**Kay M.**

*One of the best investments I ever made. Congratulations!*

Mario B.

*Happy to hear of the success. So happy I can always help! Bravo!*

Art B.

*Congratulations, LightSail 2 team and fellow Planetary Society members! We made it! LightSail 2 is beautiful!*

Russell S.

*Proud to be part of the backers. Proud to be a #PlanetarySociety member & volunteer.*

Muchinazvo S.

*I backed this when I had a tiny little boy, he's now not so tiny and very excited that we played a part (a tiny tiny tiny part) in this. This is a fantastic achievement that has inspired at least one more generation and taken us all another faltering step towards the stars and our future.*

Christopher C.

*THIS! Ordinary, Extraordinary People can Do THIS!*

Neil C.

Comments like these affirm our mission: to empower the world's citizens to advance space science and exploration. Our enthusiastic community keeps The Planetary Society moving forward, looking for more crowdfunding and citizen-action opportunities.

## How Far Can We Go?

For 40 years, The Planetary Society has been harnessing the power of the crowd both to advocate for future planetary exploration and to crowdfund private participation in those exploration efforts. The scope and complexity of our crowdfunded activities have varied and grown over time: from travel support for scientific conferences to putting payloads on Mars to funding an advanced technology mission in Earth orbit. Public support has ranged from a few thousand dollars to the $7-million LightSail project.

What is the limit for crowdfunding for space? Could a space mission costing tens of millions of dollars be crowdfunded? Could the public fund an orbiter to the Moon or Mars or fund an asteroid flyby? For projects of that size, could funds be raised based strictly on charitable motives, or will larger projects require the possibility of some financial return on investment?

Crowdfunding space projects is still a new concept, and we will need more time and more crowdfunding experiments to explore and test the boundaries. The Planetary Society is committed to this work, and by doing so is creating more—and more meaningful—opportunities for everyday people to participate in space exploration.

# References

Friedman, L. (1979). *Unpublished memorandum in the archives of the planetary society.*

Grossman, D. (July 31, 2019). LightSail 2, the solar sailing CubeSat that could, declared a success. *Popular Mechanics.* https://www.popularmechanics.com/space/satellites/a28568904/lightsail-2-declared-success/.

Lewicki, C. (May 26, 2016). *Arkyd: A space telescope for everyone. Update 39: Final update and full refund.* Kickstarter. https://www.kickstarter.com/projects/arkydforeveryone/arkyd-a-space-telescope-for-everyone-0/posts/1584844.

Manning, A. (September 29, 2017). *Bill Nye's crowdfunded lightsail can revolutionize space exploration.* Inverse https://www.inverse.com/article/36920-lightsail2-announced-bill-nye.

Sagan, C. (1980). The adventure of the planets. *The Planetary Report, 1*(1), 3—3.

The Planetary Society. (1979). *Articles of incorporation of the planetary society.* The Planetary Society. https://planetary.s3.amazonaws.com/assets/images/society/20191129_the-planetary-society-articles-of-incorporation.jpg.

# Index

Printed in the United States
by Baker & Taylor Publisher Services